T0236031

Geostatistics Explained
An Introductory Guide for Earth Scientists

This reader-friendly introduction to geostatistics provides a lifeline for students and researchers across the Earth and environmental sciences who until now have struggled with statistics. Using simple and clear explanations for both introductory and advanced material, it demystifies complex concepts and makes formulas and statistical tests easy to understand and apply.

The book begins with a discussion and critical evaluation of experimental and sampling design before moving on to explain essential concepts of probability, statistical significance and Type 1 and Type 2 error. Tests for one and two samples are presented, followed by an accessible graphical explanation of analysis of variance (ANOVA). More advanced ANOVA designs, correlation and regression, and non-parametric tests including chi-square, are then considered. Finally, it introduces the essentials of multivariate techniques such as principal components analysis, multidimensional scaling and cluster analysis, analysis of sequences (especially autocorrelation and simple regression models) and concepts of spatial analysis, including the semivariogram and its application in Kriging.

Illustrated with wide-ranging and interesting examples from topics across the Earth and environmental sciences, *Geostatistics Explained* provides a solid grounding in the basic methods, as well as serving as a bridge to more specialized and advanced analytical techniques. It can be used for an undergraduate course or for self-study and reference. Worked examples at the end of each chapter help reinforce a clear understanding of the statistical tests and their applications.

Steve McKillup is an Associate Professor in the Department of Biosystems and Resources at Central Queensland University. He has received several tertiary teaching awards, including the Vice-Chancellor's Award for Quality Teaching and a 2008 Australian Learning and Teaching Council citation "For developing a highly successful method of teaching complex physiological and statistical concepts, and embodying that method in an innovative international textbook." He is the author of *Statistics Explained: An Introductory Guide for Life Scientists* (Cambridge, 2006).

His research interests include biological control of introduced species, the ecology of soft-sediment shores and mangrove swamps.

Melinda Darby Dyar is an Associate Professor of Geology and Astronomy at Mount Holyoke College, Massachusetts. Her research interests range from innovative pedagogies and curricular materials to the characterization of planetary materials. She has studied samples from mid-ocean ridges and every continent on Earth, as well as from the lunar highlands and Mars. She is a Fellow of the Mineralogical Society of America, and the author or coauthor of more than 130 refereed journal articles. She is the author of two mineralogy DVDs used in college-level teaching, and a textbook, *Mineralogy and Optical Mineralogy* (2008).

Geostatistics Explained

An Introductory Guide for Earth Scientists

STEVE McKILLUP

Central Queensland University

MELINDA DARBY DYAR

Mount Holyoke College, Massachusetts

CAMBRIDGE
UNIVERSITY PRESS

CAMBRIDGE
UNIVERSITY PRESS

University Printing House, Cambridge CB2 8BS, United Kingdom

Cambridge University Press is part of the University of Cambridge.

It furthers the University's mission by disseminating knowledge in the pursuit of education, learning and research at the highest international levels of excellence.

www.cambridge.org
Information on this title: www.cambridge.org/9780521763226

First published 2010
Reprinted 2014

Printed in the United Kingdom by Print on Demand, World Wide

A catalog record for this publication is available from the British Library

Library of Congress Cataloguing in Publication data
McKillup, Steve.
Geostatistics explained : an introductory guide for earth scientists / Stephen McKillup, Melinda Darby Dyar.
    p.   cm.
ISBN 978-0-521-76322-6 (hardback)
1. Geology – Statistical methods.   I. Dyar, M. Darby (Melinda Darby)   II. Title.
QE33.2.S82M36   2010
550.1′5195–dc22

                                                                2010002838

ISBN 978-0-521-76322-6 Hardback
ISBN 978-0-521-74656-4 Paperback

Contents

Contents

Preface

This book presents an introduction to statistical methods that is specifically written for "earth science" students who do not have a strong background in mathematics.

The earth sciences are increasingly (and appropriately) recognized as environmental sciences that overlap and integrate with other disciplines, especially geography, hydrology, soil science, oceanography, environmental management, environmental impact assessment, bioremediation, remote sensing and conservation. As a result, the skills required of earth scientists have become far more diverse, as have the interests and backgrounds of students who enroll in these programs. Today's earth scientists need to be able to critically evaluate sampling designs, to understand the concept of statistical analysis, and be able to evaluate and interpret the results of statistical tests applied in a wide range of fields.

A sound grounding in statistical concepts and methods is especially important, but an increasing proportion of earth science students do not have this. Some have told us that math avoidance is the reason why they have pursued earth sciences instead of chemistry, biology and physics. Many such students are afraid of mathematics (often because they did badly in such subjects at high school) and dread doing an introductory statistics course.

This book has been developed for university and college courses in introductory geostatistics and as a guide for new users to learn statistics on their own. We assume very little prior knowledge of mathematics and start from first principles to develop an understanding of significance testing that can be applied to all statistical tests and related to experimental design. We use a carefully structured conceptual approach to introduce and explain what statistical tests actually do, using a minimum of terminology. Concepts that other introductory texts present as a daunting series of formulae are explained in a way that even the "math-phobic" student will find refreshing. The examples we have given are deliberately simple to help the reader understand the statistical concepts being explained. In cases where we have not given a reference for an example, the data have been deliberately contrived (or simplified from actual data) for clarity. Perhaps most importantly, this text develops a strong conceptual understanding that can be applied to the range of statistical methods used in the geosciences.

If you only take an introductory course, then this book will provide the background and understanding you need to interpret and critically evaluate results and summary reports produced by statisticians. If you go on further in geostatistics, this introduction will serve as a bridge to more advanced courses that use texts such as Borradaile (2003) *Statistics of Earth Science Data*, and Davis (1986, 2002) *Statistics and Data Analysis in Geology*.

We have many people to thank. Erick Bestland introduced us by email. Comments by reviewers improved the text. We thank our editors, Susan Francis and Jon Billam, for their considerable help and their good humor. Both our families provided enormous support and tolerated a great deal of absent-mindedness.

For Steve, Ruth McKillup provided constant encouragement and read, commented on, and reread several drafts. Lynn Stewart's constructive help was particularly appreciated, as were Haylee Weaver's insightful comments.

For Darby, thanks are due to Harold Andrews, who introduced her to statistics as an undergraduate in a course that has proven useful in many ways over the years. Tekla Harms humored many thoughtful geologic discussions at 6 a.m. Peter, Duncan and Lindy Crowley provided necessary distractions from this project and a constant reminder of what is really important. At her feet, dogs waited patiently for walks that were postponed by "one last change" to various chapters; they are glad to know that they will now have their day!

1 | Introduction

1.1 Why do earth scientists need to understand experimental design and statistics?

Earth scientists face special challenges because the things they study – the rock formations, ore bodies, deposits of minerals and fossil species – are often very large, widely dispersed and/or difficult to access. Therefore, it is usually impossible for an earth scientist to study more than a small fraction of any geological phenomenon. For example, imagine trying to measure the length of every brachiopod in the northern hemisphere, the H_2O content of every basalt flow in the USA, the diameter of every volcanic bomb on the island of Hawaii, or the orientation of every single fault plane in an entire formation. You would have to take a sample – a small subset of each – and hope that the results you obtained were representative of the larger group.

Because they are often forced to work with samples, earth scientists need to know how to sample, and they need to know how confident they can be about making generalizations from these samples.

The total number of occurrences of a particular thing (e.g. mineral species, fossil type, rock type) present in a defined area is often called the **population**. But because a researcher usually cannot measure every part of the population (unless they are studying a very restricted location, like the inside of a volcanic caldera), they have to work with a carefully selected **subset** of several **sampling units** that they hope is a **representative sample**, which can be used to infer the characteristics of the population. For example, they might measure the size (usually in terms of diagonal length) of a sample of fifty megalodon teeth from a population of several hundred, or assess the quality of a consignment of several thousand agates by breaking open a randomly chosen sample of twenty. You can also think of the population as the total number of artificial sampling units possible (e.g. all the quadrangles in the United States) and your sample being the subset

1

(e.g. 20 quadrangles) you have chosen to work with as an indication of conditions across the whole country. The concept of a representative subset also applies to experiments where you might take two (or more) samples and expose them to two (or more) different treatments. Here the replicates within each sample are often called **experimental units** to empha-size that they have been artificially manipulated. We will usually refer to replicates as sampling units in this book.

The best way to get a representative sample is usually to choose a proportion of the population at **random** – without bias, with every possible sampling unit having an equal chance of being selected.

Unfortunately it is often very difficult for earth scientists to take a random sample, because they cannot easily access the whole population. For exam-ple, it may only be possible to sample rocks that are exposed in outcrops, but these may not be the same as the rest of the formation – the outcrops may only have remained because they have a slightly different composition that makes them more resistant to weathering. A group of rocks sampled at random from float may not represent the variability present in all rocks from that outcrop/formation. Therefore, earth scientists need to know how to take the best possible sample from the part of the population they can access, and be aware of the risk of assuming that the sample is characteristic of the population.

Next, **even a random sample may not be a good representative of the population from which it has been taken.** There are often great differences among sampling units from the same population. This is not restricted to the earth sciences. Think of the people you have seen today – unless you met some identical twins (or triplets etc.), no two would have been the same. But even rock types that seem to be made up of similar-looking minerals show great variability. This leads to several problems.

First, **two samples taken at random from the same population may, simply by chance, be very different to each other and not very represen-tative of the population** (Figure 1.1).

Therefore, **if you take a random sample from each of two similar populations, the samples may be different from each other simply by chance.** On the basis of your samples, you might mistakenly conclude that the two populations are very different. You need some way of knowing if the difference between samples is what you would expect by chance, or whether the populations really do seem to be different.

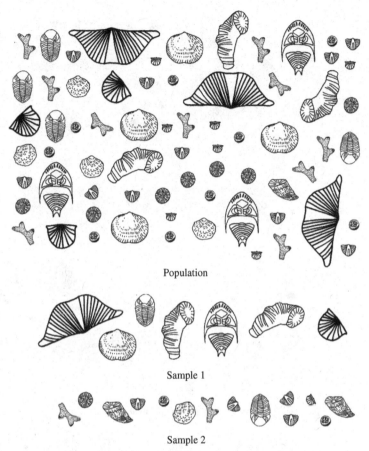

Population

Sample 1

Sample 2

Figure 1.1 Even a random sample may not necessarily be a good representative of the population. Two samples have been taken at random from a Devonian oil field in Ghawar. By chance, sample 1 contains a group of relatively large fossils, while those in sample 2 are relatively small, and the types of fossils in the two samples are also different.

Second, even if two populations are very different, samples from each may be similar simply by chance, and therefore give the misleading impression the populations are also similar (Figure 1.2).

Finally, **variation within samples may make it difficult to interpret any effect of differences in location.** There is often so much variation within a sample (and a population) that differences in location may be difficult to interpret. For example, imagine you are an environmental geologist working to assess a landfill contaminated with lead. The lead content in a sample

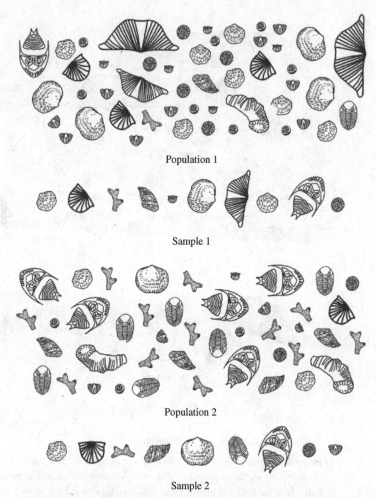

Population 1

Sample 1

Population 2

Sample 2

Figure 1.2 Samples selected at random from very different populations may not necessarily be different. Simply by chance the samples from populations 1 and 2 are similar in size and composition.

of ten cores from the oldest part of the landfill is 1000 mg/kg Pb on average, and ranges from 100–9000 mg/kg. In contrast, a sample of ten cores from the youngest part of the landfill contains 2000 mg/kg Pb on average but ranges from 100–7000 mg/kg. Which of these two areas would you consider to be most contaminated?

Variability within samples can also obscure the effect of experimental treatments. For example, opaque brown topaz crystals may change to transparent blue (which people find attractive and pay high prices for) if they are heat-treated. Gamma irradiation also alters the color of topaz. A mineralogist found that 60–80% of brown topaz crystals treated by heating turned various shades of blue. In contrast, when crystals were irradiated and then heated, a few turned bright blue, but others remained quite brown (Figure 1.3). From the extremely variable results for the 12 crystals in Figure 1.3, can you really conclude that irradiation had a significant effect?

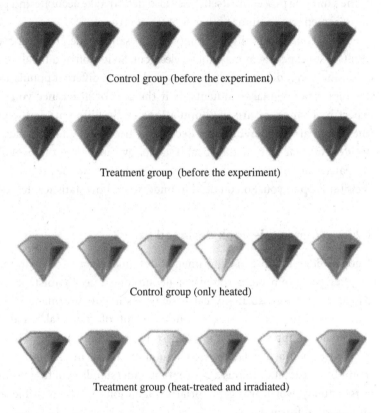

Control group (before the experiment)

Treatment group (before the experiment)

Control group (only heated)

Treatment group (heat-treated and irradiated)

Figure 1.3 Two samples of topaz crystals were taken from the same mine and deliberately matched so that six equally brown individuals were initially present in each group. Those in the treatment group were treated with ^{60}Co radiation followed by heating to 450 °C, while those in the control group were only heated. This caused all crystals to became more translucent and change color to shades of brown, pink and blue. Slightly more of the crystals in the treatment group became translucent gemmy and blue, but this difference is small compared to the variation in color among individuals, which may obscure any effect of treatment.

These sorts of problems are usually unavoidable when you work with samples and mean that a researcher has to take every possible precaution to try and ensure that their samples are likely to be **representative** and thus give a good estimate of conditions in the population. So earth scientists need to know how to sample. They also need a good understanding of experimental design, because a good sampling design will take natural variation into account and also minimize additional unwanted variability introduced by the sampling procedure itself. They also need to take accurate and precise measurements to minimize other sources of error.

Finally, considering the variability within samples described above, the results of an experiment may not be clear-cut. So it is often difficult to make a decision about differences between samples from different populations or different experimental treatments. **Is it the sort of difference you would expect by chance, or are the populations really different? Is the experimental treatment having an effect?** You need something to **help you decide**, and that is what statistical tests do, by calculating the **probability** of a particular difference among samples. Once you have the probability, the decision is up to you. So you need to understand how statistical tests work!

1.2   What is this book designed to do?

A good understanding of experimental design and statistics is important, whether you are a meteorologist, paleontologist, geochemist, seismologist or geographer, so many earth science students are made to take a general introductory statistics course. A lot of these take a detailed mathematical approach that students often find uninspiring. This book is an introduction that does not assume a strong mathematical background. Instead, it develops a conceptual understanding of how statistical tests actually work, using pictorial explanations where possible and a minimum of formulae.

If you have read other texts, or already done an introductory course, you may find that the way this material is presented is unusual, but we have found that non-statisticians find this approach very easy to understand and sometimes even entertaining. If you have a background in statistics you may find some sections a little too explanatory, but at the same time they are likely to make sense. This book most certainly will not teach you everything about the subject areas, but it will help you decide what sort of statistical test to use and what the results mean. It will also help you understand and criticize the sampling and experimental designs of others. Most importantly, it will help you design and analyze your own sampling programs and experiments, understand more complex sampling designs and move on to more advanced statistical courses

2 | "Doing science": hypotheses, experiments and disproof

2.1 Introduction

Before starting on experimental design and statistics, it is important to be familiar with how science is done. This is a summary of a very conventional view of scientific method.

2.2 Basic scientific method

The essential features of the "hypothetico-deductive" view of scientific method (see Popper, 1968) are that a person observes or samples the natural world and uses all the information available to make an intuitive logical guess, called a **hypothesis**, about it or how it functions. The person has no way of knowing if their hypothesis is correct – it may or may not apply. **Predictions** made from the hypothesis are tested, either by further sampling or by doing experiments. If the results are consistent with the predictions then the hypothesis is retained. If they are not, it is rejected, and a new hypothesis formulated (Figure 2.1). The initial hypothesis may come about as a result of observations, sampling and/or reading the scientific literature.

Here is an example. Lead contamination is an enormous environmental problem because in the past many manufacturers discarded wastes containing lead and other heavy metals into pits and landfills. These heavy metals are water soluble so they can leach into aquifers, be transported by groundwater and contaminate water supplies. In the early days, clean-up of these sites involved digging up the contaminated soil and removing it to special disposal facilities where water run-off could be contained and treated. More recently, it has been found that the mineral group apatite has a structure that easily binds to heavy metals, effectively immobilizing them. Luckily, apatite is easy to get because it is readily available in fish and mammal bones, where it is the primary constituent along with collagen.

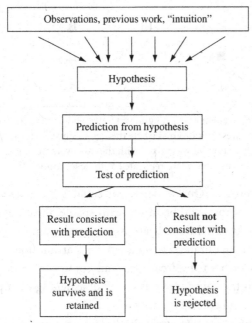

Figure 2.1 The process of hypothesis formulation and testing.

For your first remediation job as an environmental geologist, you decide to contain the lead in a contaminated landfill by mixing the soil with several tons of apatite. Your client balks at the cost, and asks you to demonstrate that it really works. The hypothesis that needs testing is simple: "Apatite will bind lead in contaminated soil."

From this hypothesis it is straightforward to predict, "Lower concentrations of lead will be present in water that has circulated through soils mixed with apatite, compared to soils without apatite."

This prediction can be convincingly tested by doing a simple and inexpensive manipulative field experiment with two treatments: (a) a 90/10 mixture of soil and apatite and (b) a 90/10 mixture of soil and an inert filler (e.g. glass beads) as a control to take into account the dilution that will occur when soil is mixed with anything else.

Because differences in the concentration of lead in the leachate might also result from heterogeneity in lead concentration across the landfill, the treatments need to be **replicated** several times. You could do this by mapping out three locations that are well spaced apart across the landfill. At each you could excavate ~20 cubic meters of soil and divide this into two

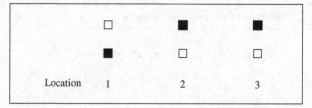

Figure 2.2 Arrangement of a 2×3 grid of treated and untreated areas in a landfill. Black squares indicate areas where the soil was mixed with apatite, and open squares where the soil was mixed with the same volume of glass beads. The treatment and its control are replicated at three locations.

equal-sized heaps (Figure 2.2). One (and here you could toss a coin to decide which) of each pair of heaps could be mixed with apatite, the other mixed with the inert glass beads, and the two heaps isolated and monitored so you could sample the water that drained from them. This arrangement would ensure that replicates of both the treatment and control were dispersed across the landfill, and the coin-tossing is a way of assigning each pair of heaps to the treatment and control at random.

You run the experiment for two weeks. Each day, you sample the water runoff from each of the six heaps, and analyze its lead content. For this manipulative experiment the three locations within each treatment are experimental units (Chapter 1).

From this experiment there are at least four possible outcomes:

(1) Run-off from the apatite-treated soil contains far lower concentrations of lead than run-off from the control. This result is consistent with the hypothesis, which has survived this initial test and can be retained.

(2) Run-off from both the apatite-treated and control soil has high concentrations of dissolved lead. This is not consistent with the hypothesis, which can probably be rejected because it seems that the apatite treatment has no effect.

(3) There is little or no dissolved lead in the run-off from either treatment. It is difficult to know if this has any bearing on the hypothesis – there may be a fault with the experiment (e.g. the $10 \, \text{m}^3$ was not enough soil, there was torrential rain during the two weeks, or maybe you did not run the experiment long enough for the rain to percolate through the heaps). The hypothesis is neither rejected nor retained.

(4) Run-off from the apatite-treated soil contains higher concentrations of lead than from the control. This is a most unexpected outcome that is

not consistent with the hypothesis, which is extremely likely to be rejected.

These are the four simplest outcomes. A more complicated and much more likely one is that you find considerable variation in lead content within **both** treatments. This sort of outcome is a problem, because you really want to keep your job! You need to figure out whether the apatite is reducing the amount of lead leached from the soil, or whether any difference between the two treatments is simply **happening by chance**. Here statistical testing is extremely useful and necessary because it helps you decide whether a difference between treatments is meaningful.

2.3   Making a decision about a hypothesis

Once you have the result of the experimental test of a hypothesis, two things can happen:

> **Either** the results of the experiment are consistent with the hypothesis, which is retained.
>
> **Or** the results are inconsistent with the hypothesis, which may be rejected.

If the hypothesis is rejected it is likely to be wrong and another will need to be proposed. If the hypothesis is retained, withstands further testing and has some very widespread generality, it may progress to become a **theory**. But a theory is only ever a very general hypothesis that has withstood repeated testing. There is always a possibility it may be disproven in the future.

2.4   Why can't a hypothesis or theory ever be proven?

No hypothesis or theory can ever be proven – one day there may be evidence that rejects it and leads to a different explanation (which can include all the successful predictions of the previous hypothesis). Consequently we can only falsify or disprove hypotheses and theories – we can never ever prove them.

Cases of disproof and a subsequent change in thinking are common. The most infamous of these in the earth sciences was the pre-twentieth-century belief that the surface of the Earth was generally similar since it was formed, with only minor changes caused by heating (expansion) and cooling (contraction) of land masses. This idea was quickly abandoned when the theory of plate tectonics, which neatly explained variations in the direction of the Earth's magnetic field as recorded in the rock record as well as fossil distributions across continents, was developed.

Another important historical example is the publication of Copernicus' famous book in 1543, which presented evidence that the stars and planets revolve around the Sun rather than the Earth. It took several decades of discussion and the invention of the telescope to make the observations that provided further support for this heliocentric perspective.

2.5   "Negative" outcomes

People are often quite disappointed if the outcome of an experiment is not what they expected and their hypothesis is rejected. But there is nothing wrong with this – rejection of a hypothesis is still progress in the process of understanding how a system functions. Therefore, a "negative" outcome that causes you to reject a cherished hypothesis is just as important as a "positive" one that causes you to retain it.

Unfortunately researchers tend to be very possessive and protective of their hypotheses, and there have been cases where results have been falsified in order to allow a hypothesis to survive. This does not advance our understanding of the world and is likely to be detected when other scientists repeat the experiments or do further experiments based on these false conclusions. There will be more about this in a later chapter on ethics, which includes discussion about doing science responsibly and ethically.

2.6   Null and alternate hypotheses

It is scientific convention that when you test a hypothesis you state it as two hypotheses which are essentially alternates. For example, the hypothesis:

"Apatite treatment reduces the amount of lead leached from soil"

is usually stated in combination with:

"Apatite treatment does **not** reduce the amount of lead leached from soil."

The latter includes all cases not covered by the first hypothesis (e.g. no difference, or more lead in leachate from the apatite treatment).

These hypotheses are called the **alternate** and **null** hypotheses respectively. Importantly, the null hypothesis is always stated as the hypothesis of "no difference" or "no effect." So, looking at the two hypotheses above, the second "does not" hypothesis is the null hypothesis and the first is the alternate hypothesis. This is a tedious but very important convention (because it clearly states the hypothesis and its alternative) and there will be several reminders in this book.

2.7   Conclusion

There are five components to an experiment – (1) formulating a hypothesis, (2) making a prediction from the hypothesis, (3) doing an experiment or sampling to test the prediction, (4) analyzing the data, and (5) deciding whether to retain or reject the hypothesis.

The description of scientific method given here is extremely simple and basic and there has been an enormous amount of philosophical debate about how science is done (see Box 2.1). For example, more than one hypothesis might explain a set of observations and it may be difficult to test these by progressively considering each one against its null. For further reading, Chalmers (1999) gives a very clearly explained discussion of the process and philosophy of scientific discovery.

Box 2.1 Two other views about scientific method

Popper's hypothetico-deductive philosophy of scientific method, where hypotheses are sequentially tested and always at risk of being rejected, is widely accepted. In reality, however, scientists may do things a little differently.

Kuhn (1970) argues that scientific enquiry does not necessarily proceed with the steady testing and survival or rejection of hypotheses. Instead, hypotheses with some generality and which have survived initial testing become well-established theories or "paradigms" which are relatively immune to rejection even if subsequent testing may find evidence against them. A few negative results are used to refine the paradigm to make it continue to fit all available evidence. It is only when the negative

---

evidence becomes overwhelming that the paradigm is rejected and replaced by a new one.

Lakatos (1978) also argues that a strict hypothetico-deductive process of scientific enquiry does not necessarily occur. Instead, fields of enquiry, called "research programmes" are based on a set of "core" theories that are rarely questioned or tested. The core is surrounded by a protective "belt" of theories and hypotheses that are tested. A successful research program is one that accumulates more and more theories that have survived testing within the belt, which provides increasing protection for the core. If, however, many of the belt theories are rejected, doubt will eventually be cast on the veracity of the core and of the research program itself, which will be replaced by a more successful one.

These two views and the hypothetico-deductive view are not irreconcilable. In all cases observations and experiments provide evidence either for or against a hypothesis or theory. In the hypothetico-deductive view science proceeds by the orderly testing and survival or rejection of individual hypotheses, while the other two views reflect the complexity of theories required to describe a research area and emphasize that it would be foolish to reject a theory outright on the basis of limited negative evidence.

2.8   Questions

(1) Describe the "hypothetico-deductive" model of how science is done, including the null and alternate hypotheses, the concepts of disproof and the importance of a negative outcome.

(2) Why is it important to collect data from more than one sampling unit or experimental unit when testing a hypothesis?

3 | Collecting and displaying data

3.1 Introduction

One way of generating hypotheses is to collect data and look for patterns. Often, however, it is difficult to see any pattern from a set of data, which may just be a list of numbers. Graphs and descriptive statistics are very useful for summarizing and displaying data in ways that may reveal patterns. This chapter describes the different types of data you are likely to encounter and discusses ways of displaying them.

3.2 Variables, sampling units and types of data

In earth science applications, we usually consider three different types of data:

(1) Data organized in a sequence along a continuum of distance or time. These data can be thought of as occurring in one dimension. For example, you might be analyzing the composition or mineralogy of a drill core and need to interpret spatial variation up and down the section.
(2) Data where sampling is done relative to some geographic or other type of spatial context. These are usually two-dimensional data. Geologic maps, contour diagrams, trend surface analyses and studies of spatial relationships in thin sections all present opportunities to relate data to a 2-D system.
(3) Multivariate data in which the 1- or 2-D locations of the sampled data are not relevant. Most types of chemical data fall into this category.

The particular attributes you measure when you collect data are called **variables** (e.g. a chemical analysis, observations of humidity and air temperature, the thickness of some geological strata). These data are collected from each sampling unit, which may be an individual (e.g. a single piece of rock) or a defined item (e.g. a square meter of the outcrop, a specific stratigraphic unit, or a particular locality).

If you only measure one variable per sampling unit the data set is **univariate**. Data for two variables per unit are **bivariate**, while data for three or more variables measured on the same sampling unit are **multivariate**.

Variables can be measured on four scales – ratio, interval, ordinal or nominal.

A **ratio scale** describes a variable whose numerical values truly indicate the quantity being measured.

- There is a true zero point below which you cannot have any data (for example, if you are measuring the length of feldspar crystals in a thin section, you cannot have a crystal of negative length).
- An increase of the same numerical amount indicates the same quantity across the range of measurements (for example, a 0.2 mm and a 2 mm feldspar will have grown by the same amount if they both increase in length by 10 mm).
- A particular ratio holds across the range of the variable (for example, a 200 μm feldspar grain is twenty times longer than a 10 μm grain and a 100 μm grain is also twenty times longer than a 5 μm one).

An **interval scale** describes a variable that can be less than zero.

- The zero point is arbitrary (for example, temperature measured in degrees Celsius has a zero point at which water freezes), so negative values are possible. The true zero point for temperature, where there is a complete absence of heat, is zero kelvin (about −273 °C), so (unlike Celsius) the kelvin is a ratio scale.
- An increase of the same numerical amount indicates the same quantity across the range of measurements (for example, a 2 °C increase indicates the same increase in heat whatever the starting temperature).
- Because the zero point is arbitrary, a particular ratio does not hold across the range of the variable. For example, the ratio of 6 °C compared to 1 °C is not the same as 60 °C to 10 °C. The two ratios in terms of the kelvin scale are 279:274 K and 333:283 K.

An **ordinal scale** applies to data where values are ranked – which means they are given a value that simply indicates their **relative order**. For example, five mountains with elevations of 10 000 m, 4500 m, 4300 m,

4000 m and 3984 m have been measured on a ratio scale. If you rank these in order, from highest to lowest, as 5, 4, 3, 2 and 1, the data have been reduced to an ordinal scale, but this is not very informative and does not mean that the highest mountain is five times the elevation of the lowest. For ordinal data, an increase in the same numerical amount of ranks does not necessarily hold across the range of the variable.

A **nominal scale** applies to data where the values are classified according to an attribute. For example, the breakdown of rocks at the Earth's surface can be classified as either chemical or mechanical weathering, so a sample of different sediments can be subdivided into the numbers within each of these two categories. You might have a sample of ten, of which three fall in the "chemical" category and the remaining seven in the "mechanical" one.

The first three types of data described above can include either **continuous** or **discrete** data. Nominal scale data (since they are attributes) can only be discrete.

Continuous data can have any value within a range. For example, any value of temperature is possible within the range from 10 °C to 20 °C, such as 15.3 °C or 17.82 °C.

Discrete data are very different from continuous data because they can only have fixed numerical values within a range. For example, the number of electrons in an atom increases from one fixed whole number to the next, because you cannot have a fraction of an electron.

It is important that you know what type of data you are dealing with because this will be one of the factors that determines your choice of statistical test.

3.3 Displaying data

A list of data may reveal very little, but a pictorial summary is a way of exploring the data that might help you notice a pattern, which can help generate or test hypotheses.

3.3.1 Histograms

Here is a list of the number of visits made to their lecturer's office by a sample of 60 students chosen at random from 320 students in the course Introductory Geoscience. These data are univariate, ratio scaled and discrete.

1, 1, 6, 1, 12, 1, 2, 6, 2, 7, 2, 2, 5, 2, 1, 2, 1, 9, 1, 8, 1, 1, 2, 5, 1, 6, 1, 1, 1, 5, 1, 1, 1, 2, 2, 3, 2, 3, 3, 3, 3, 3, 4, 5, 6, 7, 8, 9, 4, 1, 1, 9, 10, 1, 4, 10, 11, 1, 2, 3

It is difficult to see any pattern from this list of numbers, but you could summarize and display these data by drawing a histogram. To do this you separately count the number (the **frequency**) of cases for students who visited never, once, twice, three times, through to the maximum number of visits and plot these as a series of rectangles on a graph with the X axis showing the number of visits and the Y axis the number of students in each of these cases. Figure 3.1 shows a histogram of these data.

This visual summary shows that the distribution is skewed to the right – most students made few visits for help, but there is a long upper "tail" who have made five or more visits. Incidentally, looking at the graph you may be a little suspicious because every student made at least one visit. This was because each of them had to visit the lecturer's office to pick up an assignment during the first three weeks of class to ensure they knew where to go if they did ever need help, so these data are somewhat misleading in terms of indicating the neediness of the group. You may be tempted to draw a line joining the midpoints of the tops of each bar to indicate the shape of the distribution, but this implies that the data on the X axis are continuous, which is not the case because visits are discrete whole numbers.

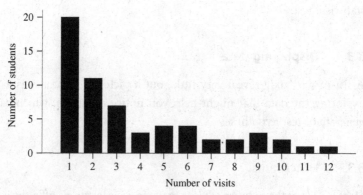

Figure 3.1 The number of visits made to their lecturer's office by a sample of 60 students chosen at random from 320 students in the course Introductory Geoscience.

3.3.2   Frequency polygons or line graphs

If the data are continuous, it is appropriate to draw a line linking the midpoint of the tops of each bar in the histogram. Here is a geological example for some continuous data that can be summarized as a histogram or as a frequency polygon (often called a line graph). Carbon isotope data are very useful for understanding the global distribution of carbon between the Earth's atmosphere, seawater and carbonate minerals. The $\delta^{13}C$ of carbonate minerals can provide information about variations of $\delta^{13}C$ in ocean water, which can be related to the global carbon cycle and palaeoceanographic circulation patterns.

A sample of 28 "muddy" limestones (wackestones) was collected from an extended outcrop, and isotopic analyses for $\delta^{13}C$ ‰ were obtained. Nothing is very obvious from this list of results:

1.01, 0.59, 2.32, 0.19, −2.39, −3.76, −0.8, 1.6, 0.28, −1.62, −0.33, −1.26, −0.01, 1.36, 0.99, 1.12, −0.45, 0.71, 1.12, −0.72, 1.36, 1.59, 2.27, 2.25, 3.05, 2.58, 1.94, 3.28

Because the data are continuous, they are not as easy to summarize as the discrete data in Figure 3.1. To display a histogram for continuous data you need to subdivide the data into the frequency of cases within a series of intervals of equal width. First you need to look at the range of the data (here $\delta^{13}C$ ‰ varies from a minimum of −3.76 through to a maximum of 3.28) and decide on an interval width that will give you an informative display of the data. Here the chosen width is 1.0 ‰. Therefore, starting from −4.0 ‰, this will give 8 intervals, the first of which is −4 to −3.01 ‰. The chosen interval width needs to be one that shows the shape of the distribution: there would be no point in choosing a width that included all the data in just two intervals because you would only have two bars on the histogram. Nor would there be any point in choosing more than 20 intervals because this would give a lot of bars with each containing only a few data.

Once you have decided on an appropriate interval size, you need to count the number of cases with $\delta^{13}C$ values that fall within each interval (Table 3.1) and plot these frequencies on the Y axis against the intervals (indicated by the midpoint of each interval) on the X axis. This has been done in Figure 3.2(a). Finally, the midpoints of the tops of each rectangle have been joined by a line to give a frequency polygon, or line graph (Figure 3.2(b)).

Table 3.1 Summary of $\delta^{13}C$ ‰ data for limestones listed as frequencies and cumulative frequencies.

Interval range $\delta^{13}C$ ‰	Cases	Cumulative Frequency	
		Total	Percent
−4 to −3.01	1	1	3.6
−3 to −2.01	1	2	7.1
−2 to −1.01	2	4	14.3
−1 to −0.01	5	9	32.1
0 to 0.99	5	14	50.0
1 to 1.99	8	22	78.6
2 to 2.99	4	26	92.9
3 to 3.99	2	28	100.0

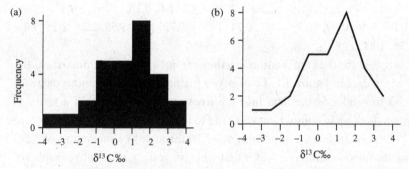

Figure 3.2 Carbon isotope data for 21 sampling units of limestone from the same outcrop, displayed as (a) a histogram and (b) a frequency polygon or line graph. The points on the frequency polygon (b) correspond to the midpoints of the bars on (a).

3.3.3   Cumulative graphs

Often it is useful to display data as a histogram of cumulative frequencies. This is a graph that displays the progressive total (starting at zero, or zero percent and finishing at the sample size or 100%) on the Y axis against the increasing value of the variable on the X axis. Figure 3.3 gives an example, using the data from Table 3.1.

A cumulative frequency graph can never decrease. Figure 3.3 displays the data in Table 3.1 as a cumulative frequency histogram.

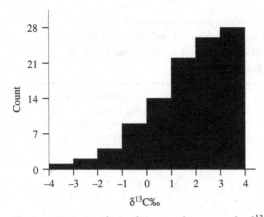

Figure 3.3 A cumulative frequency histogram for $\delta^{13}C$ data for limestones.

Although we have given the rather tedious manual procedures for constructing histograms, you will find that most statistical software packages (and spreadsheets) have excellent graphics programs for displaying your data. These will automatically select an interval width, summarize the data and plot the graph of your choice.

3.4   Displaying ordinal or nominal scale data

When you display data for ordinal or nominal scale variables, you need to modify the form of the graph slightly because the categories are unlikely to be continuous, so the bars need to be separated to clearly indicate the lack of continuity. Here is an example for some nominal scale data. Table 3.2 gives the locations of 594 tornadoes during the period from 1998–2007 in the southeastern states of the US.

These can be displayed on a bar graph with the categories in any order along the X axis and the number of cases on the Y axis (Figure 3.4(a)). It often helps to rank the data in order of magnitude to aid interpretation (Figure 3.4(b)).

3.5   Bivariate data

Data where two variables have been measured on each sampling unit can often reveal patterns that may suggest hypotheses, or be useful for testing them. Here is another case where the mineral apatite affects public health (in

Table 3.2 Preliminary data on tornado occurrence in southeastern US states from 1998–2007, according to the NOAA National Weather Service Storm Prediction Center (www.spc.noaa.gov/wcm/).

Location	Number of tornadoes 1998–2007
Texas	95
Oklahoma	68
Louisiana	38
Arkansas	68
Mississippi	68
Alabama	64
Georgia	48
Tennessee	44
North Carolina	36
South Carolina	48
Florida	17

Chapter 2 there was an example where apatite was used to clean up lead waste – this is about hydroxylapatite in your teeth). Table 3.3 gives two lists of bivariate data for the number of dental caries (these are the holes that develop in decaying teeth) and age for 20 children between the ages of one and nine years from each of the cities of Hale and Yarvard.

Looking at these data, there is not anything that stands out, apart from an increase in the number of caries with age. If you calculate descriptive statistics such as the average age and average number of dental caries for each of the two groups (Table 3.4) they are not very informative either. (You probably know how to calculate the average for a set of data and this procedure will be described in Chapter 7, but the average is the sum of all the values divided by the sample size.)

Table 3.4 shows that the sample from Yarvard had slightly more caries on average than the one from Hale, but this is not surprising because the Yarvard sample was an average of one year older. If, however, you graph these data, patterns emerge. One way of displaying bivariate data is a two-dimensional plot with increasing values of one variable on the horizontal (or X axis) and increasing values of the second variable on the vertical (or Y axis). Figure 3.5 shows both sets of data with the number of caries (Y axis) plotted against child age (X axis) for each city.

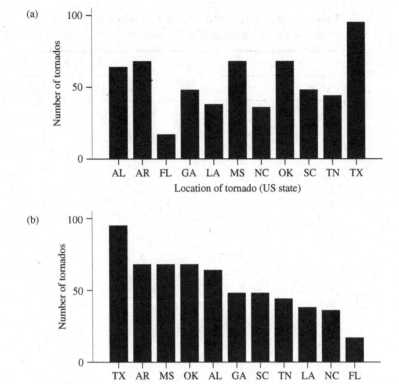

Figure 3.4 (a) Preliminary data on tornado occurrence in southeastern US states (listed alphabetically) from 1998–2007. (b) The same data but with the number of cases ranked in order from most to least.

These graphs show that tooth decay increases with age, but the pattern differs between cities – in Hale the increase is fairly steady, but in Yarvard it remains low in children up to age seven but then suddenly increases. This led to several hypotheses including that there might have been a child dental care program, or water fluoridation, in place in Yarvard for the past eight years compared to no action on decay in Hale.

Of course, there is always the possibility that the samples are different due to chance, so perhaps the first step in any further investigation would be to repeat the sampling using much larger numbers of children from each city.

Subsequent investigation found that the Yarvard municipal drinking water had been fluoridated for the past eight years, but this treatment had

Table 3.3 The number of dental caries and age of 20 children chosen at random from each of the two cities of Hale and Yarvard.

Hale		Yarvard	
Caries	Age	Caries	Age
1	3	10	9
1	2	1	5
4	4	12	9
4	3	1	2
5	6	1	2
6	5	11	9
2	3	2	3
9	9	14	9
4	5	2	6
2	1	8	9
7	8	1	1
3	4	4	7
9	8	1	1
11	9	1	5
1	2	7	8
1	4	1	7
3	7	1	6
1	1	1	4
1	1	2	6
6	5	1	2

Table 3.4 The average number of dental caries and age of 20 children chosen at random from each of the two cities of Hale and Yarvard.

Hale		Yarvard	
Caries	Age	Caries	Age
4.05	4.5 years	4.10	5.5 years

not been introduced in Hale. The fluoride program works because your teeth are made of the mineral hydroxyapatite (the same mineral that binds to heavy metals). In this case the apatite in your teeth binds fluorine ions which substitute for hydroxyls in the apatite structure, making the enamel of your teeth less soluble and therefore less prone to decay. This seems a very

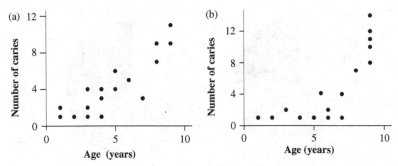

Figure 3.5 The number of dental caries plotted against the age of 20 children chosen at random from each of the two cities of (a) Hale and (b) Yarvard.

plausible reason, but bear in mind that these data are only correlative and there may be other reason(s) for the difference between the two cities.

3.6   Data expressed as proportions of a total

Data for the relative frequencies in two or more categories that sum to a total of 1.0, or 100%, can be displayed as a **pie diagram** – a circle in which each of the categories is displayed as a "slice," the size of which is proportional to its value. For example, a sample containing four different minerals that are equally abundant would be shown as a circle subdivided into four equal 90° slices. Pie diagrams are easily interpreted when there are 10 or fewer categories and each contains at least 10% of the data (Figure 3.6). When there are more than 10 categories the display will appear cluttered, especially when slices are distinguished by their color, but it will be even harder to differentiate among a lot of categories shown only as black, white and shades of grey. Categories representing a relatively small number or proportion of total cases will appear very narrow and may be overlooked.

The procedure for drawing a pie diagram showing either the relative proportion of cases in several categories, or the values of two or more variables (e.g. the concentrations of six different ions) is straightforward. First, the data for each category are listed, summed to give a total, and then expressed as proportions of this total. Each proportion is then multiplied by 360 to give the width of the slice in degrees, which is used to draw the appropriate divisions on the pie diagram.

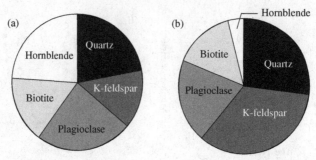

Figure 3.6 Pie diagrams comparing the mineralogy of two different granites. From this type of comparison it is clear that the rock in (a) has far less K-feldspar and much more hornblende compared to the rock in (b).

3.7   Display of geographic direction or orientation

Rose diagrams are used to show a summary of the direction or orientation of a sample of objects such as crystals or fractures in rock, or the geographic orientation of paleocurrent directions in ancient river systems. For example, a unimodal paleocurrent implies a river with steep slopes, but a bimodal one suggests a meandering river with a low slope. Rose diagrams are also commonly used by meteorologists to report the direction and magnitude of winds. The procedures for drawing rose diagrams and analyzing data for direction and orientation are described in Chapter 22.

3.8   Multivariate data

Often earth scientists have data for three or more variables measured on the same sampling unit. For example, a geologist might have data for mineralogy, chemical composition, geological age and metamorphic grade for 20 outcrops across a zone of contact metamorphism, or a paleontologist might have data for the numbers of several species of brachiopods from a specific formation.

Results for three variables could be shown as three-dimensional graphs, but direct display is difficult for more than this number of variables. Some relatively new statistical techniques have made it possible to condense and summarize multivariate data in a two-dimensional display, and these are introduced in Chapter 20.

3.9   Conclusion

Graphs may reveal patterns in data sets that are not obvious from looking at lists or calculating descriptive statistics. Graphs can also provide an easily understood visual summary of a set of results. In later chapters there will be discussion of data displays such as boxplots and probability plots, which can be used to decide whether the data set is suitable for a particular analysis. Most modern statistical software packages have easy to use graphics options that produce high-quality graphs and figures. These packages are very useful for writing assignments, reports or scientific publications.

4 | Introductory concepts of experimental design

4.1   Introduction

To generate hypotheses, you often sample different groups or locations (which is sometimes called a **mensurative** experiment because you usually measure something, such as air density, chemical composition or temperature, on each sampling unit) and explore these data for patterns or associations. To test hypotheses you may do mensurative experiments, or **manipulative** ones where you change a condition and observe the effect of that change upon each experimental unit (like the experiment with apatite and lead described in Chapter 2). Often you may do several experiments of both types to test a particular hypothesis. The quality of your sampling and the design of your experiment can have an effect on the outcome and determine whether or not your hypothesis is rejected, so it is important to have an appropriate and properly designed experiment.

First, you should attempt to make your measurements as **accurate** and **precise** as possible so they are the best estimates of actual values.

Accuracy   is the closeness of a measured value to the true value.
Precision   is the "spread" or variability of repeated measures of the same value.

For example, a thermometer that consistently gives a reading corresponding to a true temperature (e.g. 20 °C) is both accurate and precise. Another that gives a reading consistently higher (e.g. +10 °C) than a true temperature is not accurate, but it is very precise. In contrast, a thermometer that gives a fluctuating reading within a wide range of values around a true temperature is not precise and will usually be inaccurate except when the reading occasionally happens to correspond to the true temperature.

The distinction between accuracy and precision is particularly important (and sometimes quite vexing) in geochemical studies. Most types of analytical instruments are calibrated relative to standards (usually glasses or minerals), but even many "true" or "standard" values for the compositions of naturally occurring glasses or minerals have themselves been obtained using analytical techniques that are subject to error. So an important (though rather prosaic) part of geochemistry is the characterization of standards. This is usually done by analyzing a standard in several different ways, repeating these in different laboratories and using the combined results to determine an accepted "true" value for it (that is usually an average). Because geoscientists deal with naturally occurring materials such approximations are often unavoidable and the errors they contribute to analytical work are an underlying source of inaccuracy in most types of geochemical data.

Inaccurate and imprecise measurements or a poor or unrealistic sampling design can result in the generation of inappropriate hypotheses. Measurement errors or a poor experimental design can give a false or misleading outcome that may result in the incorrect retention or rejection of an hypothesis.

The following is a discussion of some important essentials of sampling and experimental design.

4.2   Sampling: mensurative experiments

Mensurative experiments are often a good way of generating or testing predictions from hypotheses. (An example of the latter is "I think apatite binds heavy metals. So if I sample groundwater at 500 sites with high concentrations of naturally occurring apatite and 500 where apatite concentrations are low or zero, the groundwater in the first group should, on average, contain less dissolved heavy metals.") You have to be careful when interpreting the results of mensurative experiments because you are sampling an **existing condition**, rather than **manipulating** conditions experimentally. There may be some other difference between your groups (e.g. the "high apatite" sites may have negligible amounts of heavy metals present anyway).

4.2.1   Confusing a correlation with causality

A correlation between two variables means they vary together. A positive correlation means that high values of one variable are associated with high values of the other, while a negative correlation means that high values of one variable are associated with low values of the other. For example, the graph in Figure 4.1 shows a positive correlation between the increasing tilt of the summit of a volcano and the number of earthquakes measured there. A volcano with a flat summit has zero tilt.

Unfortunately a correlation is often mistakenly interpreted as indicating causality. In this example it seems very plausible that the tilt of the summit could be caused by settling that occurs as a result of the earthquakes, so volcanoes with more frequent earthquakes are more likely to have tilted summits. However, even if there is a very obvious correlation between any two variables, it does not necessarily show that one is responsible for the other. The correlation may have occurred by chance, or a third unmeasured factor might determine the values of the two variables studied. In this case, tilting actually occurs because of increased pressure from fresh magma moving into the chamber beneath the summit, which inflates the ground beneath the summit, both increasing the tilt and causing rock fracturing that is manifested as earthquakes. There is no causal relationship between earthquakes and tilt: they are both caused by the volume of magma injected into the chamber (Figure 4.2).

Another complication in the interpretation of correlation arises often in geological analyses involving **closed data sets**, where the variables **sum to a fixed total, percentage or proportion** such as 100% or 1.0. This type of error is particularly common in reports of chemical analyses of rocks and minerals. For example, summary data for mineral formulae are routinely

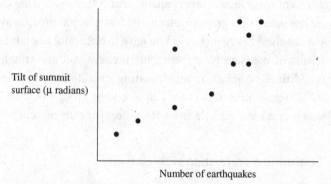

Tilt of summit
surface (μ radians)

Number of earthquakes

Figure 4.1 Example of a positive correlation between the tilt of the summit of a volcano and the number of earthquakes observed there.

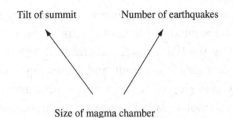

Figure 4.2 The involvement of a third variable "size of magma chamber" that determines the tilt of the volcano's summit and the number of earthquakes that occur. Even though there is no causal relationship between tilt and the number of earthquakes, these two variables are positively correlated.

summed to the appropriate number of cations. The two olivine minerals fayalite (Fe_2SiO_4) and forsterite (Mg_2SiO_4) rarely occur in these two "pure" forms. Instead, most specimens contain a mixture of iron and magnesium, each of which can fit in the two available octahedral sites in the structure, which is why you might see the formula for olivine written as $Fe_{0.6}Mg_{1.4}SiO_4$. The cations Fe and Mg must sum to two, so they will always be inversely correlated with each other. An increase in Fe will inevitably be accompanied by a decrease in the percentage of Mg, but this does not mean one has an effect on the other. Rather, the formula probably reflects the amounts of Mg and Fe in the original melt when the mineral crystallized. Care must be taken to avoid being misled by such spurious negative correlations in closed systems.

4.2.2 The inadvertent inclusion of a third variable: sampling confounded in time

Occasionally researchers have no choice but to sample different sites at different times. These results should be interpreted with great caution, because changes occurring over time may contribute to differences (or the lack of them) among samples or sampling units. The sampling is said to be **confounded** in that more than one variable may be having an effect on the results. Here is an example.

An oceanographer measured the sodium content of carbonate shelf sediments that are being formed in the Holyoke estuary, by taking 100 cores running in a line down the length of the estuary. Coring is time-consuming and the oceanographer only had access to a boat for one day per week and could only take five cores per day. Therefore the most upstream part of the estuary was sampled in May, the middle in June and the most downstream sites in August. The sampling showed a positive correlation between the sodium content of newly laid down sediment and increasing distance downstream. Unfortunately, however, the change in sodium content was actually caused by seasonal variation in salinity that was low throughout the entire estuary in spring due to groundwater run-off and increased as the year progressed (and the scientist did not know this). Subsequent work where the entire estuary was sampled on the same day found no difference in the sodium content of the sediments, so the correlation between distance downstream and sodium content was an artifact of the sampling of different places being confounded in time. This is an example of a common problem, and you are likely to find similar cases in many published scientific papers and reports.

4.2.3   The need for independent samples in mensurative experiments

Frequently researchers have to accurately describe relatively large areas or objects as part of a mensurative experiment. For example, annual sedimentation rates in glacier-fed lakes are used to reconstruct up-valley glacier activity and thereby model environmental change. Sediment traps are deployed at various locations in these lakes, and the mean particle size, quantity and sediment flux are calculated. There is an obvious need to **replicate** the sampling – that is, to independently measure sedimentation rate in more than one place.

If you only sampled sediment in one trap at one place (Figure 4.3(a)) the results would not be a good indication of the sediment accreting across the whole lake. The sampling needs to be replicated, but there is little value in repeatedly sampling one small area (e.g. by taking several samples under "*****" in Figure 4.3(b)) because this still will not give an accurate indication of variation in sediment rate and composition across the whole lake (although it may give a very accurate indication of conditions in that particular part of the lake sampled). This sort of sampling is one aspect of what Hurlbert (1984) called **pseudoreplication**, which is still a very common flaw in a lot of scientific research. The replicates are "pseudo" – sham or unreal – because they are unlikely to truly describe what is occurring

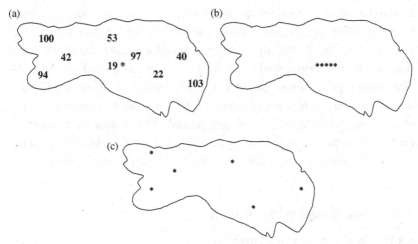

Figure 4.3 Aerial view of a glacial lake showing mean sediment grain size (in µm). (a) An unreplicated sample taken at only one place (*) would give a very misleading indication of grain size within the entire lake. (b) Several replicates taken close to one another (*****) would still give a very misleading indication. (c) Several replicates taken at random across the lake would give a better indication.

across the entire area being discussed (in this case the lake). A better design would be to sample at several places chosen at random within the lake, as shown in Figure 4.3(c).

Here is another example. A researcher sampled a peatland ecosystem by dropping a $10\,m^2$ square frame, subdivided into a grid of 100 equal sized squares, at random in one place only and then took one sample from each of these smaller squares. Although these 100 replicates may very accurately describe conditions within the sampling frame, they do not necessarily describe the remaining $9990\,m^2$ of the wetland environment and would be pseudo-replicates if the results were interpreted in this way. A more appropriate design would be to sample 100 replicates chosen at random across the entire area.

4.2.4   The need to repeat the sampling on several occasions and elsewhere

The results of sampling systems which are in considerable flux (e.g. the Holyoke estuary or the lake described above) can only confidently be discussed in relation to the area at the time of sampling. Therefore, you need to be cautious when interpreting results. Sampling the same area over several different years will strengthen the findings, and may be sufficient if you are only interested in that area. Sampling more than one area (e.g. estuary, lake or outcrop) will make the results more able to be generalized. Inappropriate generalization is another example of pseudoreplication since data from one location (whether it is relatively unchanging or in a state of flux) may not hold in the more general case. At the same time, however, even if your study is limited, you can still make more general predictions from your findings provided these are clearly identified as predictions.

4.3   Manipulative experiments

4.3.1   Independent replicates

It is essential to have several independent replicates of any treatment used in a manipulative experiment. We mentioned this briefly when describing the lead remediation experiment in Chapter 2 and said that for only one treated and one untreated sample, any difference between them could have simply been due to chance or some other unknown factor(s). As the number of randomly chosen independent replicates increases, so does the likelihood that any difference between the experimental group and the control group is a result of the experimental treatment.

Here is an example of the need for replicates. Streams that drain from catchments where gold is present often contain particles of alluvial gold among the gravel on the stream bed. This gold can be recovered by "panning" – placing a few handfuls of stream gravel and sediment in the base of a shallow pan, part-filling the pan with water and then gently swirling the contents to wash away any fine sediment, after which the gravel is spread out within the pan and the gold grains picked out. A lot of experience is needed before a person is skilled at doing this, and amateurs miss a lot of gold. The standard gold pan is a wide and shallow metal bowl. One day you notice that an older gold pan which has rusted and pitted on the inside seems to give a more even spread of gravel, making it much easier to see the gold. To test the hypothesis that a pitted pan gives better recovery you buy two new standard gold pans, hammer 1000 pits into one and take both to a gold bearing stream where you pan 16 ounces of gravel with the pitted one while a friend pans another 16 ounces with the standard. To your delight you recover 0.025 ounces of gold, compared to 0.007 ounces for the standard one. The result is obvious – the pitted pan is better.

This result is consistent with the hypothesis but there is an obvious flaw in the experiment – with only one pan in each treatment, any difference between them may be due to differences between the pans, the amount of gold initially present in the gravel sample, the skill of the operators (including that you might have concentrated more carefully on the contents of the pitted pan because you hoped to recover more gold, or your friend did a poor job out of disinterest), or all three. There is a need to replicate this experiment and the replicates need to be truly independent – it is not sufficient for you to use the pitted pan ten times while your friend does the same with the standard pan, because any differences between treatments may still be caused by operator skill. There will be more about this shortly.

4.3.2   Control treatments

Control treatments are needed because they allow the experimenter to isolate the reason why something is occurring in an experiment by comparing two treatments that differ by only **one factor**. Frequently the need for a rigorous experimental design makes it necessary to have several different treatments, more than one of which can be considered controls.

Here is an example. Ultramafic rocks, and the soils produced from their weathering, naturally contain high concentrations of metals such as chromium, cobalt, manganese and nickel (that are toxic to plants) and low concentrations of elements needed for plant growth. Therefore, very few plant species grow in ultramafic soils and the ones that do are sometimes so distinctive that they can be mapped by remote sensing and used to identify sites with particularly high concentrations of valuable metals, especially nickel, for mining.

Some of these plant species also accumulate extremely high concentrations of nickel in their leaves and stems, and it has recently been suggested that these accumulators could be grown on contaminated land and the mature plants later harvested and removed (and perhaps even smelted), to reduce soil contamination. There is great interest in such **bioremediation** and a environmental scientist did an experiment to trial it on a nickel-contaminated mine tailing. The soil was plowed and planted with 10 000 seedlings of a nickel accumulator and the plants were left to grow for six months. At the end of the experiment the result was clear – less nickel was present in the soil at the treatment site.

Unfortunately this experiment is not controlled. By plowing the site and planting a nickel accumulator, more than one treatment has been applied. Furthermore, six months have elapsed between measurements. So if you do find the nickel content of the soil has decreased, it may have been the result of any of these factors and you cannot confidently attribute it to the presence of the nickel accumulator.

One essential improvement would be to have a control for the presence of the plant, where a site was plowed but not planted, and then left for the same amount of time. Another might simply be a control for time, where nothing was done to the site. At this stage, by incorporating these two improvements, you would have three treatments. Table 4.1 lists what these treatments are doing to the tailings site.

For this design, if there is a reduction in nickel in the treatment with the plants and no reduction in the other two, you might feel confident in claiming the nickel accumulator had an effect. Nevertheless, some mine site rehabilitators might say the design is still inadequate because the treatment in the left-hand column of Table 4.1 is the only one in which **any** plant has been grown. For example, the roots of the plants may simply have made the soil more permeable to rainwater which leached some of the nickel, and the same result might have been obtained by planting 10 000 seedlings of an inexpensive non-accumulating species. This improvement has been listed in Table 4.2.

At this point you may be thinking that the above design is far too complex, but experiments do have to have appropriate controls so that the effects of each potentially contributing factor can be isolated. For this example, it would be extremely embarrassing to spend several thousand dollars on expensive seedlings only to find that **any** plant has the same effect.

Table 4.1 Breakdown of three treatments in terms of their effect upon a nickel-rich site.

	Control for disturbance	Control for time
Planted nickel accumulator		
Nickel accumulator		
Plowing	Plowing	
Time	Time	Time

Table 4.2 Breakdown of three treatments in terms of their effect upon a nickel-rich site.

Planted nickel accumulator	Control for plant	Control for disturbance	Control for time
Nickel accumulator	Non accumulator		
Plowing	Plowing	Plowing	
Time	Time	Time	Time

It is often difficult to work out what control treatments you need for an experiment. One way to clarify these is to list all of the things that are actually happening in an experimental treatment and make sure you have appropriate controls for each. Finally, the experimental treatments need to be appropriately replicated – you cannot just have one replicate of each of the four treatments in Table 4.2.

Here you may be wondering about the applicability of these concepts of experimental design to geological systems, where you yourself do not (and cannot) manipulate the experiment *per se*, because the "treatment" might be some process or event that happened millions of years ago. Often earth scientists can only do mensurative experiments, but a knowledge of the essentials (and pitfalls) of doing these is becoming increasingly necessary if you work in areas such as mine and hazardous waste site remediation.

4.3.3 Pseudoreplication

One of the nastiest pitfalls is appearing to have a replicated manipulative experimental design that really is not replicated. This is another aspect of "pseudoreplication" described by Hurlbert (1984) who invented the word – before then it was just called "bad design." Here is an example that relates back to the use of bioaccumulators to rehabilitate mine sites.

A remediation geoscientist hypothesized that simply plowing the soil would help decontaminate a mine site. They were aware of the need to have appropriate treatments and replicates, so they chose two separate tailing sites each 1000 m² in area. A toss of a coin decided which of these was plowed, and the other site was left undisturbed. One hundred replicate experimental units were taken at random within each site and analyzed for nickel, which was high and similar at both. After six months the nickel content in the soil was resampled and found to be much lower at the plowed site. The scientist was delighted – an inexpensive experiment with 100 replicates of each treatment had produced a result consistent with the hypothesis.

Unfortunately, there are **not** 100 truly independent replicates in each area because the treated site is in a different place to the control. All replicates of the plowed treatment were in one tailing and all those from the undisturbed one were in another. Therefore, any difference in nickel may, or may not, have been due to the plowing – it could equally well have been due to some other (perhaps unknown) difference between the sites. The number of replicates is the sites, not the number of experimental units within each, so the experiment has no effective replication at all and is essentially the same as the (unreplicated) manipulative experiment on gold panning described earlier in this chapter.

An improvement to the design would be to run each of the two treatments at several tailing sites, but here too, it is still be necessary to have truly independent replicates. If you do not, the experiment may still suffer from **apparent replication**, and here are four examples.

(1) Treatments separate but clumped. Even if you have several separate replicates of each treatment, the arrangement of these can lead to a lack of independence. For example, you may have your treatments all clumped together at one end of the mining lease and the controls at the other, but this is no better than an unreplicated example (Figure 4.4(a)).

(2) Replicates placed alternately. If you decided to get around the clustering problem by placing treatments and controls alternately (i.e. by placing, from east to west, treatment #1, control #1, treatment #2, control #2, treatment #3 etc.) there can still be problems. Just by chance all the treatment sites (or all the controls) might be in line with an underlying feature of the area (e.g. a regular alternation of grain size or soil composition), or subject to some other regular feature you are not even aware of (Figure 4.4(b)).

(3) Unavoidable segregation of replicates. Often, due to practical considerations, you have to have all of your replicates of one treatment in only one place, and all replicates of the control group in another. Unfortunately, if there is something peculiar to one location, in addition to the variable you are intentionally manipulating, then either the

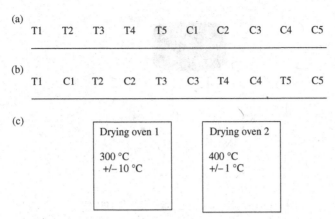

(a)

| T1 | T2 | T3 | T4 | T5 | C1 | C2 | C3 | C4 | C5 |

(b)

| T1 | C1 | T2 | C2 | T3 | C3 | T4 | C4 | T5 | C5 |

(c)

Drying oven 1

300 °C
+/– 10 °C

Drying oven 2

400 °C
+/– 1 °C

Figure 4.4 Three cases of apparent pseudoreplication. (a) Clustering of replicates means that there is no independence among controls or treatments. (b) A regular arrangement of treatments and controls may, by chance, correspond to some feature of the environment that might affect the results. (c) Segregation of treatments within particular ovens, where perhaps the variance in temperature might be different.

experimental or control treatment may be affected. For example, if you were doing an experiment comparing crystallization at two high temperatures you might have access to only two ovens, one set at 300 °C and the other at 400 °C. Unfortunately these ovens may differ in more ways than their set temperature. One may vary in temperature by +/– 10 °C, while the other might be more accurate and only vary by only +/– 1 °C. This pattern is called "isolative segregation" (Figure 4.4(c)).

(4) The final example is more subtle. Imagine you are an aqueous geochemist interested in the hypothesis that "pH affects the minerals that crystallize from sulfate-rich water." You set up five control beakers and five experimental beakers and place them on the bench in a completely randomized pattern to get around problems in (a), (b) and (c) above. All the beakers have water constantly flowing through them, so you set up two storage tanks, one with water of pH 3.0 and the other with a higher pH of 7.0. Water from each storage tank is piped into five beakers as shown in Figure 4.5.

This looks fine, but unfortunately all five beakers within each treatment are sharing the same water. All in the "pH 7" treatment receive water from Tank A and all beakers in the "pH 3" treatment receive water from Tank B,

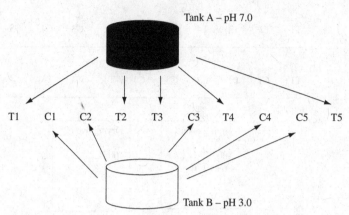

Figure 4.5 The positions of the treatment beakers are randomized, but all tanks within a treatment share water from one supply tank.

so any difference in precipitated minerals between treatments may be due either to the pH or some other feature of the supply tanks and circulation system. Really, therefore, this design is little better than the case of isolative segregation (example (c) above). Ideally, each beaker should have its own separate and independent supply. Finally, the allocation of replicate beakers to treatments should be done using a method that removes any possibility of unintentional bias by the experimenter. (For example, the toss of a coin was used to allocate paired heaps of soil to treated and untreated areas in the experiment with lead and apatite described in Section 2.2.)

4.4   Sometimes you can only do an unreplicated experiment

Although replication is desirable in any experiment, there are some cases where it is not possible. For example, when doing large-scale mensurative or manipulative experiments on systems such as lakes or rivers, there may be only one polluted lake or river available to study. Although you cannot attribute the reason for any difference, or the lack of it, to the treatment (e.g. a polluted versus a relatively unpolluted river) because you only have one replicate of each, the results are still useful. First, they are still evidence for or against your hypothesis and can be cautiously discussed in the light of the lack of replication. Second, it may be possible to achieve replication by analyzing your results in conjunction with those from similar studies done elsewhere by other researchers. This is called a **meta-analysis**. Finally, the results of a large-scale but unreplicated experiment may suggest smaller-scale experiments that can be done with replication so that you can continue to test the hypothesis.

4.5   Realism

Even an apparently well-designed mensurative or manipulative experiment may still suffer from a lack of realism. Here are two examples.

The first is a mensurative experiment on the occurrence of impact craters on the Earth. It was once thought that meteor impacts on Earth were fairly rare, and the contrast between the appearance of the Moon (which is pockmarked with craters), and the Earth's surface (which is almost lacking in visible craters), supported the hypothesis "Impacts rarely occur on the surface of the Earth compared to the Moon." The null hypothesis (that few people even considered to be a possibility) was that impacts occur at about the same frequency on the surface of the Earth and the Moon.

All this changed with two big events during the 1960s. First, the return of samples from manned lunar exploration revealed the relative ages of different sizes of lunar craters and supported the idea that impacts are occurring continuously, although with a decreasing frequency, in our solar system. Second, the theory of plate tectonics was introduced, which said that new crust is constantly being created and subducted all over the Earth's surface. Suddenly it was apparent that the huge impact basins on the lunar surface were created more than four billion years ago, and if comparable craters existed on the Earth we would have no record of them. In the following decades, studies of what is now known as Shoemaker Crater in Arizona identified the characteristics of terrestrial impact craters (shocked minerals, enrichment of trace elements only found in meteorites, etc.). Now, numerous examples of impact craters have been found, particularly in older terrains like the Canadian shield. So the naive assumption that impacts never occur on the surface of the Earth turned out to be quite uninformed and the comparison was unrealistic because surfaces of vastly different ages on the Moon and Earth were being compared.

Second, a geoscientist did an experiment to test the hypothesis that the amount of olivine (Mg_2SiO_4) produced when MgO and SiO_2 react together is dependent on temperature. They used equilibration temperatures of 700, 800, 900 and 1000 °C, and heated the mixtures for one week, after which the treatment samples were removed from the furnaces and inspected for olivine growth. The conclusion was that there was no effect of temperature on olivine production. Later, it was realized that these temperatures were unrealistically low compared to those occurring within the Earth's crust and a subsequent experiment that included replicates at 1300, 1400 and 1500 °C showed more olivine grew at the higher temperatures.

4.6 A bit of common sense

By now, you may be quite daunted by the challenge of being able to design a good experiment. Provided, however, that you have appropriate controls, replicates and have also thought about any obvious problems of pseudo-replication and realism, you are well on the way to a good design. Furthermore, the desire for a near-perfect design has to be balanced against financial constraints as well as space and time available to do the experiment, so often it is not possible to have as many replicates as you would like. It also depends on the type of science you do. For example, if you were doing a precisely controlled standardization using several different methods you would be unlikely to set up different replicates of each treatment in a random pattern within the laboratory and to do this might grossly increase the risk of making a procedural error (not to mention the cost of the experiment!). Similarly, most experimental petrologists working with gas-mixing furnaces are very careful to calibrate thermostats, identify temperature gradients and locate furnace hotspots accurately, so that conditions can be strictly controlled. They would never be concerned about clustering of replicates or isolative segregation because they were confident that conditions did not vary among furnaces or in different parts of the same one. Most of the time they may be right, but considerations about experimental design need to be borne in mind by all scientists, especially if you are working in areas where conditions cannot be strictly maintained.

Sometimes you may not have the resources to do a large manipulative field experiment at more than one site. Certainly, in many geological studies, there may well be only one accessible road-cut or outcrop suitable for study. Although, strictly speaking, the individual results cannot be generalized to other sites, they may nevertheless apply and with careful interpretation and discussion of results you can make more general predictions. For example, the "apatite remediation" experiment described in Chapter 2 was initially conceived in the laboratory, and repeated at numerous test sites before being approved by the US Environmental Protection Agency. All the results were consistent with the hypothesis, so the general consensus at present is that "Apatite treatment reduced the amount of lead available for leaching." Nevertheless, the hypothesis may not be correct or apply to all lead remediation sites, but, to date, there has been no evidence to the contrary. Furthermore, the hypothesis is also supported by sound scientific arguments for why the treatment works: the lead has a divalent charge (Pb^{2+}) and it substitutes readily for Ca^{2+} in the apatite ($\sim Ca_5(PO_4)_3(F,Cl,OH)$) structure because it has a similar size and the same charge.

4.7 Designing a "good" experiment

Designing a well-controlled, appropriately replicated and realistic experiment has been described by some researchers as an "art." It is not, but there are often several different ways to test the same hypothesis, hence several different experiments that could be done. Consequently, it is difficult to set a guide to designing experiments beyond an awareness of the general principles discussed in this chapter.

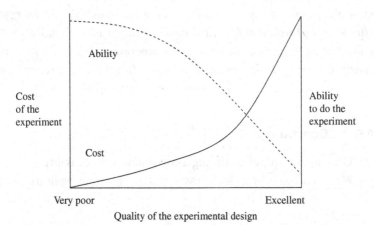

Figure 4.6 An example of the trade-off between the cost and ability to do an experiment. As the quality of the experimental design increases, so does the cost of the experiment (solid line), while the ability to do the experiment decreases (dashed line). Your design usually has to be a compromise between one that is practicable, affordable and of sufficient rigor. Its quality can be anywhere along the X axis.

4.7.1   Good design versus the ability to do the experiment

It has often been said "There is no such thing as a perfect experiment." One inherent problem is that as a design gets better and better, the cost in time and equipment also increases, but the ability to actually do the experiment decreases (Figure 4.6). An absolutely perfect design may be impossible to carry out. Therefore, every researcher must choose a design that is "good enough" but still practical. This trade-off is illustrated in Figure 4.6. The quality of an experiment can be any point along the X axis and the "best" compromise is not necessarily where the two lines cross – instead the decision on design quality is in the hands of the researcher, and will be eventually judged by their colleagues who examine any report from the work.

4.8   Conclusion

The above discussion only superficially covers some important aspects of experimental design. Considering how easy it is to make a mistake, you probably will not be surprised that a lot of published scientific papers have serious flaws in design or interpretation that could have been avoided. Work with major problems in the design of experiments is still being done and, quite alarmingly, many researchers are not aware of these. As an example, after teaching the material in this chapter, we often ask our students to find a published paper, review and criticize the experimental design, and then offer constructive suggestions for improvement. Many have later reported that it was far easier to find a flawed paper than they expected.

4.9   Questions

(1) Give an example of confusing a correlation with causality.
(2) Name and give examples of two types of "apparent replication."

5 | Doing science responsibly and ethically

5.1 Introduction

By now you are likely to have a very clear idea about how science is done. Science is the process of rational enquiry, which seeks explanations for natural phenomena. Scientific method was discussed in a very prescriptive way in Chapter 2 as the proposal of a hypothesis from which predictions are made and tested by doing experiments. Depending on the results, which may have to be analyzed statistically, the decision is made to either retain or reject the hypothesis. This process of **knowledge by disproof** advances our understanding of the natural world and seems impartial and hard to fault.

Unfortunately, this is not necessarily the case because **science is done by human beings who sometimes do not behave responsibly or ethically**. For example, some scientists fail to give credit to those who have helped propose a new hypothesis. Others make up, change or delete results so their hypothesis is not rejected, omit details to prevent the detection of poor experimental design, and deal unfairly with the work of others. Most scientists are not taught about responsible behavior and are supposed to learn a code of conduct by example. Considering the number of cases of scientific irresponsibility that have been exposed, this does not seem to be a very good strategy. Thus, this chapter is about the importance of behaving responsibly and ethically when doing science.

5.2 Dealing fairly with other people's work

5.2.1 Plagiarism

Plagiarism is the theft and use of techniques, data, words or ideas without appropriate acknowledgment. If you are using an experimental technique or procedure devised by someone else, or data owned by another person, you must acknowledge this. If you have been reading another person's work, it is easy to inadvertently use some of their phrases, but plagiarism is the repeated and excessive use of text without acknowledgment. Once your work is published, any detected plagiarism can affect your credibility and career. Quite remarkably, we have detected plagiarism in manuscripts we have been asked to review, including cases where material from the same journal has been copied.

5.2.2   Acknowledging previous work

Previous studies can be extremely valuable because they can add weight to a hypothesis and even suggest other hypotheses to test. There is a surprising tendency for scientists to fail to acknowledge previous published work by others in the same area, sometimes to the extent that experiments done two or three decades ago are repeated and presented as new findings. This can be an honest mistake in that the researcher is unaware of previous work, but it is now far easier to search the scientific literature than it used to be. When you submit your work to a scientific journal for publication, it may be embarrassing to be told that something similar has been done before. Even if a reviewer or the editor of a journal does not notice, others may and are likely to say so in print.

5.2.3   Fair dealing

Some researchers cite the work done by others in the same field but downplay or even distort it. Although it appears that previous work in the field has been acknowledged because the publication is listed in the citations at the back of the paper or report, the researcher has nevertheless been somewhat dishonest. We have found this in about five percent of the papers we have reviewed, but it may be more common because it is quite hard to detect unless you are very familiar with the work. Often the problem seems to arise because the writer has only read the abstract of a paper, which can be misleading. It is important to carefully read and critically evaluate previous work in your field because it will improve the quality of your own research.

5.2.4   Acknowledging the input of others

Often hypotheses may arise from discussions with colleagues or with your supervisor. This is an accepted aspect of how science is done. If, however, the discussion has been relatively one-sided in that someone has suggested a useful and novel hypothesis to you, then you should seriously think about acknowledgment. A colleague once said bitterly "My suggestions become someone else's original thoughts in a matter of seconds." Acknowledgment can be a mention (in a section headed "Acknowledgments") at the end of a report or paper, or you may even consider including the person as an author. If you are ever in doubt as to which of these is appropriate, remember that being generous and including someone as a coauthor (with his/her permission, of course) costs you very little, compared with the risk of alienating them if they are omitted. Often asking the person what they think is appropriate will solve this problem.

It is not surprising that disputes often arise between supervisors and their postgraduate students about authorship of papers. Some supervisors argue that they have facilitated all of the student's work by being the supervisor and therefore expect their name to be included on all papers from the research. Others recognize that single-authored papers may be important to the student's future, and thus do not insist on this. The decision depends on the amount and type of input and rests with the principal author of the paper, but it is often helpful to clarify the matter of authorship and acknowledgment with your supervisor(s) at the start of a postgraduate program or new job.

5.3   Doing the sampling or the experiment

5.3.1   Approval

In some cases, you will need prior permission to undertake an experiment, including submitting a risk assessment for an experimental procedure or field-work that may expose you, or others, to potential hazards. If you are sampling in a national park or reserve, you will need a permit. In both cases, you will have to give a well-reasoned argument for doing the work, including its likely advantages and disadvantages. In many countries and institutions, there are severe penalties for breaches of permits or doing research without prior permission.

5.3.2   Ethics

Ethics are moral judgments where you have to decide if something is right or wrong, so different scientists can have different ethical views. Ethical issues include honesty and fair dealing, but they also extend to whether experimental procedures can be justified. For example, some scientists think it is right to test cosmetic products on animals such as rabbits or rats because it will reduce the likelihood of harming or causing pain to humans, while others think it is wrong because it may cause pain and suffering to the animals. Both groups of scientists would probably be puzzled if someone said it was unethical to do experiments on insects or plants. Similarly, some scientists believe it is wrong to extract minerals or oil from areas of wilderness because of the potential for damaging these ecosystems, while others believe the need to obtain these resources is sufficient justification for extraction. Importantly, however, none of these views can be considered the best or most appropriate, because ethical standards are not absolute. Provided a person honestly believes, for any reason, that it is right to do what he is doing, then he is behaving ethically (Singer, 1992) and it is up to you to decide what is right. The remainder of this section is about the ethical conduct of research, rather than whether a research topic or procedure is considered ethical.

5.4   Evaluating and reporting results

Once you have the results of an experiment, then you need to analyze them and discuss the results in terms of rejection or retention of your hypothesis. Unfortunately, some scientists have been known to change the results of experiments to make them consistent with their hypothesis, which is grossly dishonest. We suspect this practice is more common than reported; it may even be encouraged by assessment procedures in universities and colleges where grades are given for the correct outcomes of practical experiments. When we ask undergraduate students in our statistics classes if they have ever altered their data to fit the expectations of their assignments, we tend to get a lot of very guilty looks. We have also known researchers who were dishonest. One had a regression line that was not statistically significant, so they changed the data until it was. The second made up entire sets of data for sampling on field trips that never occurred, and a third made up large quantities of data for the results of laboratory analyses that were queried by their supervisor because the data were "too good." All were found out and are no longer doing science.

It has been suggested that part of the problem stems from people becoming attached to their hypotheses and believing they are true, which goes completely against science proceeding by disproof! Some researchers are quite downcast when their results are inconsistent with their hypothesis. However, you need to be impartial about the results of any experiment and remember that a negative result is just as important as a positive one because the understanding of the world has progressed in both cases.

Another cause of dishonesty is that scientists are often under extraordinary pressure to provide evidence for a particular hypothesis. There are often career (and financial) rewards for finding solutions to problems or suggesting new models of natural processes. Competition among scientists for jobs, promotion and recognition is intense and can also foster dishonesty.

The problem with scientific dishonesty is that the person has not reported what is really occurring. Science aims to describe the real world, so if you fail to reject a hypothesis when a result suggests you should, you will report a false and misleading view of the process under investigation. Future hypotheses and research based on these findings are likely to produce results inconsistent with your findings. There have been some spectacular cases where scientific dishonesty has been revealed, which have only served to undermine the credibility of the scientific process.

5.4.1   Pressure from peers or superiors

Sometimes inexperienced, young or contract researchers have been pressured by their superiors to falsify or give a misleading interpretation of their results. It is far better to be honest than risk being associated with work that may subsequently be shown to be flawed. One strategy for avoiding such pressure is to keep good written records.

5.4.2   Record keeping

Some research groups, especially in industry, are so concerned about honesty that they have a code of conduct: all researchers have to keep records of their ideas, hypotheses, methods and results in a hard-bound laboratory book with numbered pages that are signed and dated on a daily or weekly basis by the researcher and supervisor. Not only can this be scrutinized if there is any doubt about the work (including who thought of something first), but it also encourages good data management and sequential record keeping. Results kept on pieces of loose paper with no reference to the methods used can be quite hard to interpret when the work is written up for publication.

5.5 Quality control in science

Publication in refereed journals ensures your work is scrutinized by at least one referee who is a specialist in the research field. Nevertheless, this process is more likely to detect obvious and inadvertent mistakes than deliberate dishonesty and many journal editors have admitted that work they publish is likely to be flawed (LaFollette, 1992). Institutional strategies for quality control of the scientific process are becoming more common and many have rules about the storage and scrutiny of data. At the same time, however, there is a need in many institutions for explicit guidelines about the penalties for misconduct, together with mechanisms for handling alleged cases of misconduct reported by others. The responsibility for doing good science is often left to the researcher. It applies to every aspect of the scientific process, including devising logical hypotheses, doing well-designed experiments and using and interpreting statistics appropriately, together with honesty, responsible and ethical behavior, and fair dealing.

5.6 Questions

(1) A college lecturer said "For the course 'Geostatistical Methods,' the grade a student gets for the exam has always been fairly similar to the one they get for the assignment, give or take about 15%. I am a busy person, so I will simply copy the assignment grades into the column marked 'exam' on the spreadsheet and not bother to grade the exams at all. It's really fair of me, because students get stressed during the exam anyway and may not perform as well as they should." Please discuss.

(2) An environmental scientist said "I did a small pilot experiment with two replicates in each treatment and got the result we hoped for. I didn't have time to do a bigger experiment but that didn't matter – if you get the result you want with a small experiment, the same thing will happen if you run it with many more replicates. So when I published the result, I said I used twelve replicates in each treatment." Please comment thoroughly and carefully on all aspects of this statement.

6 | Probability helps you make a decision about your results

6.1   Introduction

Most science is comparative. Earth scientists often need to know if a particular phenomenon has had an effect, or if there are differences in a particular variable measured at several different locations. For example, what is the permeability of sandstone with and without carbonate impurities? How does turbidity vary across a glacial lake? How well does the distribution of dew point temperature predict rainfall? But when you make these sorts of comparisons, any differences among areas sampled or manipulative experimental treatments may be real or they may simply be the sort of variation that occurs by chance among samples from the same population.

Here is an example of commercial importance. Most diamonds are mined from kimberlite deposits, which are volcanoes that have risen from great depths in the Earth's mantle at high speed. Sometimes, the kimberlite brings along diamonds that have formed at high pressures and temperatures. But not all kimberlites contain diamonds, and finding them within these rocks is quite difficult.

Fortunately, many kimberlites contain large amounts of the mineral garnet. A prospector noted that the garnets present in diamond-rich kimberlites were slightly darker than those in kimberlites lacking diamonds, and subsequent research suggested that the change in color was caused by the presence of small amounts of oxidized Fe, or Fe^{3+}. To test if oxidized garnets could be used to predict the presence of diamonds, the prospector collected 14 garnet samples: seven from diamond-bearing deposits of kimberlite, and seven from kimberlite without diamonds (Table 6.1), and measured their Fe^{3+} content.

On average the %Fe^{3+} content of garnets from diamond-bearing deposits is 2.8% higher than those from diamond-free deposits, but by looking at the

Table 6.1 The %Fe^{3+} content of garnets with and without coexisting diamonds.

Sample	Without diamond	With diamond
1	1.0	1.5
2	0.5	3.3
3	0.7	5.8
4	2.3	3.2
5	1.1	6.7
6	0.8	4.2
7	1.4	2.5
Average	1.1	3.9

data you can see that there is a lot of variation within both groups and even some overlap between them.

Even so, the prospector might conclude that diamonds are associated with oxidized (Fe^{3+}-bearing) garnets. But there is a problem. How do you know that this difference between groups is **meaningful** or **significant?** Perhaps it simply occurred by chance and the oxidation state of the garnet is not a good predictor? Somehow you need a way of helping you **make a decision** about your results.

Even when there may seem to be a sound scientific explanation for the phenomenon you observe, statistics can be very useful in making a decision about your results. The need to make such decisions led to the development of tests that provide a commonly agreed-upon level of statistical significance.

6.2   Statistical tests and significance levels

Statistical tests are just a way of working out **the probability of obtaining the observed, or an even more extreme, difference among samples (or between an observed and expected value) if a specific hypothesis (usually the null of no difference) is true**. Once the probability is known, the experimenter can make a decision about the difference, using criteria that are uniformly used and understood. Here is a very easy example where the probability of every possible outcome can be calculated.

Imagine you visit the beach after a big storm, and notice that the usually white sand has now turned gray. What has happened is that many of the lighter-colored minerals (feldspar and quartz) have washed out to sea, leaving behind a larger than usual percentage of black grains (the amphibole group mineral hornblende). You do not know this, but the proportions of white and black grains in this population on the beach are exactly 1 : 1. The grains are well mixed and all have exactly the same, well-rounded, shape. They are a population of many billion grains of sand.

You take one grain at random from the beach. Because there are equal numbers of black and white, your probability of getting a black one is 50%, or 1/2, which is also your chance of getting a white one. The chance of getting **either** a black or white grain is the sum of these probabilities: (1/2 + 1/2) which is 1.0 (or 100%) since there are no other colors. (If you are unsure about probability, there is a short explanation of the concepts you will need for this book in Box 6.1.)

Now consider what happens if you take a sample of six grains from the beach in sequence, one after the other, without looking. (The population is so large that removing only six will have a negligible effect on the remainder, so these are independent events: see Box 6.1.)

Box 6.1 Essential concepts of probability

The probability of any event can only vary between 0 and 1 (which correspond to 0 and 100%). If an event is **certain to occur** it has a probability of 1, while if an event is certain **not** to occur it has a probability of 0.

The probability of a particular event is the number of outcomes giving that event, divided by the total number of possible outcomes. For example, when you toss a coin, there are only two possible outcomes – a head or a tail. These two events are mutually exclusive – you cannot get both. Consequently, the probability of a head is 1 divided by 2 = 1/2 (and thus the probability of a tail is also 1/2).

Probability is usually symbolized as P, so the sentence above could be written as P (head) = 1/2 and P (tail) = 1/2.

The addition rule

The probability of getting **either** a head **or** a tail is the sum of the two probabilities, which is 1/2 + 1/2 = 1, or P (head) + P (tail) = 1. This is an example of the **addition rule**: when several outcomes are **mutually exclusive** (meaning they cannot occur simultaneously), the probability of getting any of these is the sum of their separate probabilities. (Therefore, the probability of getting a 1, 2, 3 or 4 when rolling a six-sided die is 4/6.)

The multiplication rule

Independent events. When the occurrence of one event has no effect on the occurrence of the second, the events are independent. For example, if you tossed two coins simultaneously, the outcome (H or T) for the first coin would have no influence on the outcome for the second (and vice versa). To calculate the joint probability of two or more independent events such as two heads occurring when two coins are tossed simultaneously, which would be written as P (head, head), you simply multiply the independent probabilities together. Therefore, the probability of getting two heads with two coins is P (head) $\times$ P (head) which is $1/2 \times 1/2 = 1/4$. The chance of a head **or** a tail with two coins is 1/2 because there are two ways of obtaining this out of the four possible outcomes: coin 1 = H, coin 2 = T or vice versa.

Related events. If the events are not independent (for example, for a single roll of a six-sided die, the first event being a number in the range of 1–3 inclusive, and the second event being that this is an even number) the multiplication rule also applies, but you have to multiply the probability of one event by the **conditional probability** of the second.

When rolling a die the independent probability of a number from 1–3 is $3/6 = 1/2$, and the independent probability of any even number is also 1/2 (the even numbers are 2, 4 or 6 divided by the six possible outcomes).

If, however, you have already rolled a number from 1–3, the probability of that restricted set of outcomes being an even number is 1/3 (because "2" is the only even number possible in this set of three outcomes). Therefore, the probability of **both** related events is $1/2 \times 1/3 = 1/6$. You can work out this probability the other way – the chance of an even number when rolling a die is 1/2 (you would get numbers 2, 4 or 6) and the probability of one of these numbers being in the range from 1–3 is 1/3 (the number 2 out of these three outcomes). Therefore the probability of both is again is $1/2 \times 1/3 = 1/6$.

First event	Second event		Product
(a) Even number	Number from 1–3, *provided* first event is an even number		
$P = 1/2$	$P = 1/3$		1/6
(b) Number from 1–3	Even number, *provided* first event is a number from 1–3		
$P = 1/2$	$P = 1/3$		1/6

The conditional probability of an event (e.g. an even number provided a number from 1–3 has already been rolled) occurring is written as $P(A|B)$ which means "the probability of event A *provided* event B has occurred." For the example with the die, the probability of an even number, provided a number from 1–3 has been rolled, is written as $P(\text{even}|1\text{–}3)$.

Here are all of the possible outcomes. You may get six black grains or six white ones (both outcomes are very unlikely); five black and one white, or one black and five white (which are more likely); four black and two white, or two black and four white (which are even more likely), or three black and three white (which is very likely because the proportion of grains on the beach is 1:1).

The probability of getting six black grains in sequence is the probability of getting one black one (1/2) multiplied by itself six times, which is $1/2 \times 1/2 \times 1/2 \times 1/2 \times 1/2 \times 1/2 = 1/64$.

The probability of getting six white grains is also 1/64.

The probability of five black and one white is greater because there are six ways of getting this combination (WBBBBB or BWBBBB or BBWBBB or BBBWBB or BBBBWB or BBBBBW) giving 6/64.

There is the same probability (6/64) of getting five white and one black.

The probability of four black and two white is even greater because there are 15 ways of getting this combination (WWBBBB, BWWBBB, BBWWBB, BBBWWB, BBBBWW, WBWBBB, WBBWBB, WBBBWB, WBBBBW, BWBWBB, BWBBWB, BWBBBW, BBWBWB, BBWBBW, BBBWBW) giving 15/64.

There is the same probability (15/64) of getting four white and two black.

Finally, the probability of three black and three white (there are 20 ways of getting this combination) is 20/64.

You can summarize all of these outcomes as a table of probabilities (Table 6.2). These probabilities are shown as a histogram in Figure 6.1. Note that the distribution is symmetrical with a peak corresponding to the cases where half the grains will be black and half white. (Incidentally, this is an example of the **binomial distribution,** which will be discussed in Chapter 7.)

Table 6.2 The probabilities of obtaining all possible combinations of black and white grains in samples of six from a large population where there are equal numbers of black and white grains.

Number of black	Number of white	Probability of this outcome	Percentage of cases likely to give this result
6	0	1/64	1.56
5	1	6/64	9.38
4	2	15/64	23.44
3	3	20/64	31.25
2	4	15/64	23.44
1	5	6/64	9.38
0	6	1/64	1.56
	Total:	64/64	100%

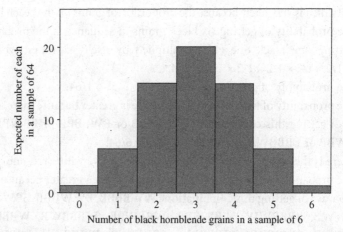

Figure 6.1 The expected numbers of each possible mixture of colors when sampling six grains independently with replacement on 64 different occasions from a large population containing 50% black hornblende and 50% white quartz grains.

Therefore, if you were given an extremely large population containing 50% black hornblende and 50% white quartz grains, from which you drew six, you would have a very high probability of drawing a sample that contains grains of both minerals. It is very unlikely you would get only six black or six white (the probability of each is 1/64, so the probability of **either** six black or six white is the sum of these which is only 2/64, or 0.0313 or 3.13%).

6.3   What has this got to do with making a decision or statistical testing?

The statistician Sir Ronald Fisher proposed that if the probability of getting the observed difference, plus any more extreme than this, between the expected outcome (the null hypothesis discussed in Chapter 2) and the actual outcome is **less than 5%**, then it is appropriate to conclude that the difference is **statistically significant** (Fisher, 1954).

There is no scientific reason for the choice of 5% (which is the same as 1/20 or 0.05). It is the probability that many researchers use as a standard "statistically significant level."

Using the example of the grains on the beach, if your null hypothesis specified that there were equal numbers of hornblende and quartz grains in the population, then you could do an experiment to test it by taking a random sample of six grains as described above. If the six grains were all black (hornblende), then the probability of this result under the null hypothesis would be only 1.56%. Similarly, if all six grains were white, the probability under the null hypothesis would also be 1.56%. Therefore, for either outcome, the difference between the experimental outcome and the expected result has such a low probability that it would be considered statistically significant. A researcher would reject the null hypothesis and conclude that the sample did **not** come from a population containing equal numbers of hornblende and quartz grains.

6.4   Making the wrong decision

If the proportions of black and white grains on the beach really were equal, then most of the time a sample of six grains would contain both minerals. But if the grains in the sample were all only black hornblende or all only white quartz, a researcher would decide the population did not contain 50% hornblende and 50% quartz. Here they would have made the wrong decision but this would not happen very often (the probability of either of these outcomes is 2/64).

The unavoidable problem with using probability to help you make a decision is that there is **always** a chance of making a wrong decision and you have no way of telling when you have done this.

As described above, if a researcher got a sample of six of one mineral, they would decide that the population on the beach was not 50% hornblende and 50% quartz when really it was. This mistake, where the null hypothesis of equal numbers is inappropriately rejected, is called a **Type 1 error**.

There is another problem. Sometimes an unknown population is different to what is expected (e.g. it may contain 90% white grains and 10% black ones) but the sample taken (e.g. 4 white and 2 black) is not significantly different to the expected outcome predicted by the hypothesis of 50:50. In this case the researcher would decide the composition of the population was the one expected under the null hypothesis (50:50), even though it was not. This mistake, when the alternate hypothesis holds but is inappropriately rejected, is called a **Type 2 error**.

Every time you do a statistical test you run the risk of a Type 1 or Type 2 error. There will be more discussion of these errors in Chapter 9, but they are unavoidably associated with using probability to help you make a decision.

6.5   Other probability levels

Sometimes, depending on the hypothesis being tested, a researcher may decide that the "less than 5%" significance level (with its 5% chance of inappropriately rejecting the null hypothesis) is too risky.

Here is an example from medical mineralogy. Mesothelioma is a cancer of the pleural mesothelium (the lining of the lung cavity), and it is mainly caused by exposure to asbestos fibers. Asbestiform minerals take the shape of fibers with longitudinal parting, and the ends then fray into individual fibers. If inhaled, the fibers are either coughed up or remain in the lung where they become covered with white blood cells called macrophages that engulf foreign particles. Unfortunately, asbestos fibers are difficult to dislodge by coughing, and their large surface area makes it difficult for macrophages to engulf them. The resultant inflammation and scarring of lung tissue leads to a high incidence of cancer.

A mineralogist helped a drug company develop and test a new and extremely expensive drug that was hoped to reduce mortality in people suffering from mesothelioma. A large experiment was done where half of mesothelioma cases chosen at random received the new drug and the other half did not. The survival of both groups over the next month was compared. The alternate hypothesis was "There will be increased survival of the drug-treated group compared to the control."

Here, the prohibitive cost of the drug meant that the manufacturer had to be very confident that it was of real use before recommending and marketing it. Therefore, the risk of a Type 1 error (significantly greater survival in the experimental group compared to the control simply by chance) when using the 5% significance level might be considered too risky. Instead, the researcher might decide to reduce the risk of Type 1 error by using the 1% level and only recommend the drug if the reduction in mortality was so marked that it was significant at this level.

Here is an example of the opposite case. A company developed a new and extremely economical method for measuring the concentration of arsenic in groundwater. Here, the company had to be extremely confident that their new method gave readings that did not differ significantly from the established method, so two thousand samples were analyzed using both. The null hypothesis was that "The estimated concentration of arsenic does not differ between methods." Here a real difference that went undetected in the trial could be disastrous for public health, so the company statistician used a 30% significance level to reduce the risk of getting a non-significant difference due to chance.

The most commonly used significance level is 5%, which is 0.05. If you decide to use a different level in an analysis, the decision needs to be made, justified and clearly specified before the sampling or the experiment is done.

For a significant result, the actual probability is also important. For example, a probability of 0.04 is not very much less than 0.05. In contrast, a probability of 0.002 is very much less than 0.05. Therefore, even though both are significant, the result with the lowest probability gives much stronger evidence for rejecting the null hypothesis.

6.6   How are probability values reported?

The symbol used for the chosen significance level (e.g. 0.05) is the Greek α (alpha). Often you will see the probability reported as $P < 0.05$ or $P < 0.01$ or $P < 0.001$. These mean respectively "The probability is less than 0.05" or "The probability is less than 0.01" or "The probability is less than 0.001." N.S. means "not significant," which is when the probability is 0.05 or more ($P \geq 0.05$). Of course, as noted above, if you have specified a significance level of 0.05 and get a result with a probability of less than 0.001, this is far stronger evidence for your alternate hypothesis than a result with a probability of 0.04.

6.7   All statistical tests do the same basic thing

In the "grains of sand" example all of the possible outcomes were listed and the probability of each was calculated directly. Some statistical tests do this. Most, however, use a formula to produce a number called a **statistic**. The probability of getting each possible value of the statistic has been previously calculated so you can use the formula to get the numerical value of the statistic, look up the probability of that value in a published set of statistical tables and make your decision to retain the null hypothesis if it has a probability of ≥ 0.05, or reject it if it has a probability of < 0.05. Most statistical software packages now available will generate the probability as well as the statistic, so you do not even need a set of tables.

6.8   A very simple example: the chi-square test for goodness of fit

Here is an example to illustrate the concepts discussed above, using one of the simplest statistical tests.

The chi-square test for goodness of fit compares observed ratios to expected ratios for nominal scale data. Imagine you have developed a new method for treating zircons to turn them from brown into colorless, gemmy ones. After doing many experiments, you predict that the success rate of your technique should be 3:1 brown:colorless. Therefore, when you irradiate 100 brown zircons from a newly discovered locality you would expect the brown:colorless ratio in the samples to be 75:25. (Your null hypothesis is

---

Box 6.2 Bayes' theorem

The calculation of the probability of two events by multiplying the probability of the first by the conditional probability of the second in Box 6.1 is an example of **Bayes' theorem**. Put formally, the probability of events A and B occurring, is the probability of event B multiplied by the probability A will occur **provided** event B has already occurred:

$$P(A, B) = P(B) \times P(A|B)$$

As described in Box 6.1, the probability of an even number **and** a number from 1–3 in a single roll of a die: $P(\text{even}, 1\text{–}3) = P(1\text{–}3) \times P(\text{even}|1\text{–}3)$.

Here is an example of the use of Bayes' theorem. In central Queensland many rural property owners have a well drilled in the hope of accessing underground water, but there is a risk of not striking sufficient water (i.e. a maximum flow rate of less than 100 gallons per hour is considered insufficient) and there is also a risk that the water is unsuitable for human consumption (i.e. it is not potable). It would be very helpful to know the probability of the combination of events of **striking sufficient water that is also potable:** $P(\text{sufficient, potable})$.

Obtaining $P(\text{sufficient})$ is easy, because drilling companies keep data for the numbers of sufficient and insufficient wells they have drilled. Unfortunately they do not have records of whether the water is potable, because that is established later by a laboratory analysis paid for by the property owner. Furthermore, laboratory analyses of samples from new wells are usually only done on those that yield sufficient water – there would be little point of assessing the water quality of an insufficient well. Therefore, data from laboratory analyses for potability only gives the conditional probability $P(\text{potable}|\text{sufficient})$. Nevertheless, from the two known probabilities, the chance of striking sufficient and portable water can be calculated:

$$P(\text{sufficient, potable}) = P(\text{sufficient}) \times P(\text{potable}|\text{sufficient}).$$

From drilling company records the likelihood of striking sufficient water in central Queensland (P sufficient) is 0.95 (so it is not surprising that one company charges 5% more than its competitors but guarantees to refund the drilling fee for any well that does not strike sufficient water).

Laboratory records for water sample analyses show that only 0.3 of sufficient wells yield potable water (P potable|sufficient).

Therefore, the probability of the two events **sufficient and potable water** is only 0.285, which means that the chance of this occurring is slightly more than 1/4. If you were a central Queensland property owner with a choice of two equally expensive alternatives of (a) installing additional rainwater tanks, or (b) having a well drilled, what would you decide on the basis of this probability?

The outcome of two events A and B occurring together, $P(A,B)$, can be obtained in two ways:

$$P(A, B) = P(B) \times P(A|B) = P(A) \times P(B|A)$$

Here too, this formula can be used to obtain probabilities that cannot be obtained directly. For example, by rearrangement the conditional probability of $P(A|B)$ is:

$$P(A|B) = \frac{P(A) \times P(B|A)}{P(B)}$$

This has widespread applications that are covered in more advanced texts.

that "The ratio of brown to colorless is no different from 3:1.") When the treatment was applied it produced 86:14 brown:colorless, which is somewhat less successful than your prediction. This might be due to chance, it may be because your null hypothesis is incorrect, or a combination of both. You need to decide whether this result is significantly different from the one expected under the null hypothesis.

This is the same as the concept developed in Section 6.2 when we discussed sampling sand grains on a beach, except that the chi-square test for goodness of fit generates a statistic (a number) that allows you to easily estimate the probability of the observed (or any greater) deviation from the expected outcome. It is so simple you can do it on a calculator.

To calculate the value of chi-square, which is symbolized by the Greek χ^2, you take each expected value away from its equivalent observed value, square the difference and divide this by the expected value. These separate values (two in the case above) are added together to give the chi-square statistic.

First, here is the chi-square statistic for an expected ratio that is the same as the observed (observed numbers 75 brown : 25 colorless; expected 75 brown : 25 colorless). Therefore the two categories of data are "brown" and "colorless."

$$\chi^2 = \frac{(75 - 75)^2}{75} + \frac{(25 - 25)^2}{25} = 0 + 0 = 0$$

The value of chi-square is zero when there is no difference between the observed and expected values.

As the difference between the observed and expected values increases, so does the value of chi-square. Here the observed ratio is 74 and 26. The value of chi-square can only be a positive number because you always square the difference between the observed and expected values.

$$\chi^2 = \frac{(74 - 75)^2}{75} + \frac{(26 - 25)^2}{25} = 0.0533$$

For an observed ratio of 70:30, the chi-square statistic is:

$$\chi^2 = \frac{(70 - 75)^2}{75} + \frac{(30 - 25)^2}{25} = 1.333$$

When you take samples from a population in a "category" experiment you are, by chance, unlikely to always get perfect agreement to the ratio in the population. For example, even when the ratio in the population is 75:25, some samples will have that ratio, but you are also likely to get 76:24, 74:26, 77:23, 73:27 etc. The range of possible outcomes among 100 samples goes all the way from 0:100 to 100:0. So the distribution of the chi-square statistic generated by taking samples in two categories from a population in which there really is a ratio of 75:25 will look like the one in Figure 6.2, and the most unlikely 5% of outcomes will generate values of the statistic that will be greater than a critical value determined by the number of independent categories in the analysis.

Going back to the result of the gemstone treatment experiment given above, the expected numbers are 75 and 25 and the observed numbers are 86 brown and 14 colorless.

To get the value of chi-square value, you calculate:

$$\chi^2 = \frac{(86 - 75)^2}{75} + \frac{(14 - 25)^2}{25} = 6.453$$

The critical 5% value of chi-square for an analysis of two independent categories is 3.841. This means that only the most extreme 5% of departures from the expected ratio will generate a chi-square statistic **greater than this**

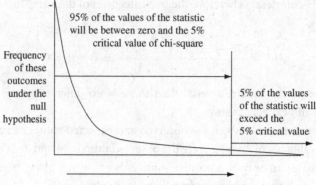

95% of the values of the statistic will be between zero and the 5% critical value of chi-square

Frequency of these outcomes under the null hypothesis

5% of the values of the statistic will exceed the 5% critical value

Increasingly positive value of chi-square

Figure 6.2 The distribution of the chi-square statistic generated by taking samples from a population containing only two categories in a known ratio. Many of the samples will have the same ratio as the expected and thus generate a chi-square statistic of zero, but the remainder will differ from this by chance, thus giving positive values of chi-square. The most extreme 5% departures from the expected ratio will generate statistics greater than the critical value of chi-square.

value. There will be more about the chi-square test in Chapter 18, including reference to a table of critical values in Appendix A.

Because the actual value of chi-square is 6.453, the observed result is significantly different to the result expected under the null hypothesis. The researcher would conclude that the ratio in the population sampled is not 3:1 and therefore reject the null hypothesis. It sounds like your new gemstone treatment is not as good as predicted (because only 14% were transformed compared to the expected 25%), so you might have to revise your estimated success rate of converting brown zircons into colorless ones.

6.9   What if you get a statistic with a probability of exactly 0.05?

Many statistics texts do not mention this and students often ask "What if you get a probability of **exactly 0.05?**" Here the result would be considered **not significant** since significance has been defined as a probability of **less than 0.05** (< 0.05). Some texts define a significant result as one where the probability is **less than or equal to 0.05** (≤ 0.05). In practice this will make very little difference, but since Fisher proposed the "less than 0.05" definition, which is also used by most scientific publications, it will be used here.

More importantly, many researchers would be uneasy about any result with a probability close to 0.05 and would be likely to repeat the experiment because it is so close to the critical value. If the null hypothesis applies then there is a 0.95 probability of a non-significant result on any trial, so you would be unlikely to get a similarly marginal result when you repeated the experiment.

6.10   Conclusion

All statistical tests are a way of obtaining the probability of a particular outcome. This probability is either generated directly as shown in the "grains from a beach" example, or a test that generates a statistic (e.g. the chi-square test) is applied to the data. A test statistic is just a number that usually increases as the difference between an observed and expected value (or between samples) also increases. As the value of the statistic becomes larger and larger, the probability of an event generating that statistic gets smaller and smaller. Once the probability of that event or one more extreme is less than 5%, it is concluded that the outcome is statistically significant.

A range of tests will be covered in the rest of this book, but most of them are really just methods for obtaining the probability of an outcome that helps you make a decision about your hypothesis. Nevertheless, it is important to realize that the probability of the result does not make a decision for you, and that even a statistically significant result may not necessarily have any geological significance – the result has to be considered in relation to the system you are investigating.

6.11   Questions

(1) Why would many scientists be uneasy about a probability of 0.06 for the result of a statistical test?
(2) Define a Type 1 error and a Type 2 error.
(3) Discuss the use of the 0.05 significance level in terms of assessing the outcome of hypothesis testing. When might you use the 0.01 significance level instead?

7 | Working from samples: data, populations and statistics

7.1 Using a sample to infer the characteristics of a population

Usually you cannot study the whole population, so every time you gather data from a sample you are "working in the dark" because the sample may not be very representative of that population. You have to take every possible precaution, including having a good sampling design, to try to ensure a representative sample. Unfortunately you still do not know whether it **is** representative! Although it is dangerous to extrapolate to the more general case from measurements on a subset of individuals, that is what researchers have to do whenever they cannot work on the entire population.

This chapter discusses statistical methods for estimating the characteristics of a population from a sample and explains how these estimates can be used for significance testing.

7.2 Statistical tests

Statistical tests can be divided into two groups, called **parametric** and **non-parametric** tests. Parametric tests make certain assumptions, including that the data fit a known distribution. In most cases this is a **normal distribution** (see below). These tests are used for ratio, interval or ordinal scale variables. Non-parametric tests do not make so many assumptions. There is a wide range of non-parametric tests available for ratio, interval, ordinal or nominal scale variables.

7.3 The normal distribution

A lot of variables, including "geological" ones, tend to be normally distributed. For example, if you measure the slopes of the sides of 100 cinder cones

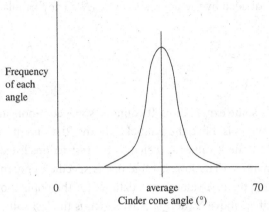

Figure 7.1 An example of a normally distributed population. The shape of the distribution is symmetrical about the average and the majority of values are close to the average, with an upper and lower "tail" of steeply and gently sloping cinder cones, respectively.

chosen at random and plot the frequency of these on the Y axis and angle on the X axis, the distribution will look like a symmetrical bell, which has been called the **normal distribution** (Figure 7.1).

The normal distribution has been found to apply to many types of variables in natural phenomena (e.g. grain size distributions in rocks, the shell length of many species of marine snails, stellar masses, the distribution of minerals on beaches, etc.).

The very useful thing about normally distributed variables is that two **descriptive statistics** – the **mean** and the **standard deviation** – can describe this distribution. From these, you can predict the proportion of data that will be less than or greater than a particular value. Consequently, tests that use the properties of the normal distribution are straightforward, powerful and easy to apply. To use them you have to be sure your data are reasonably "normal." (There are methods to assess normality and these will be described later.)

To understand parametric tests you need to be familiar with some statistics used to describe the normal distribution and some of its properties.

7.3.1   The mean of a normally distributed population

First, the mean (the average) symbolized by the Greek μ describes the location of the center of the normal distribution. It is the sum of all the values (X_1, X_2 etc) divided by the population size (N). The formula for the mean is:

$$\mu = \frac{\sum\limits_{i=1}^{N} X_i}{N} \; . \tag{7.1}$$

This formula needs some explanation. It contains some common standard abbreviations and symbols. First, the symbol Σ means "the sum of" and the symbol X_i means "All the X values specified by the restrictions listed below and above the Σ symbol." The lowest value of i is specified underneath Σ (here it is 1, meaning the first value in the data set for the population) and the highest is specified above Σ (here it is N, which is the last value in the data set for the population). The horizontal line means that the quantity above this line is divided by the quantity below. Therefore, you add up all the values (X_1 to X_N) and then divide this number by the size of the population (N).

(Some textbooks use Y instead of X. From Chapter 3 you will recall that some data can be expressed as two-dimensional graphs with an X and Y axis. Here we will use X and show distributions with a mean on the X axis, but later in this book you will meet cases of data that can be thought of as values of Y with distributions on the Y axis.)

As a quick example of the calculation of a mean, here is a population of only four fossil snails ($N = 4$). The shell lengths in mm of these four individuals (X_1 through to X_4) are 6, 7, 9 and 10, so the mean, μ, is $32 \div 4 = 8$ mm.

7.3.2   The variance of a population

The mean describes the location of the center of the normal distribution, but two populations can have the same mean but very different dispersions around their means. For example, a population of four snail fossils with shell lengths of 1, 2, 9 and 10 mm will have the same mean, but greater dispersion, than another population of four with shell lengths of 5, 5, 6 and 6 mm.

There are several ways of indicating dispersion. The **range**, which is just the difference between the lowest and highest value in the population, is sometimes used. However, the **variance**, symbolized by the Greek σ^2, provides a lot of information about the normal distribution that can be used in statistical tests.

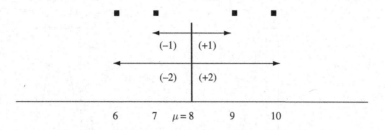

Differences squared: 4   1   1   4
Sum of the squared differences = 10
Population size = 4
Population variance = (10 ÷ 4) = 2.5

Figure 7.2 Calculation of the variance of a population consisting of only four fossil snails with shell lengths of 6, 7, 9 and 10 mm, each indicated by the symbol ■. The vertical line shows the mean μ. Horizontal arrows show the difference between each value and the mean. The numbers in brackets are the magnitude of each difference, and the contents of the box show these differences squared, their sum and the variance obtained by dividing the sum of the squared differences by the population size.

To calculate the variance, you first calculate μ. Then, by subtraction, you calculate the difference between each value $(X_1...X_N)$ and μ, square these differences (to convert each to a positive quantity) and add them together to get the sum of the squares, which is then divided by the sample size. This is similar to the way the average is calculated, but here you have **an average value for the dispersion**.

This procedure is shown pictorially in Figure 7.2 for the population of only four snail fossils, with shell lengths of 6, 7, 9 and 10 cm.

The formula for the above procedure is straightforward:

$$\sigma^2 = \frac{\sum_{i=1}^{N}(X_i - \mu)^2}{N} \qquad (7.2)$$

If there is no dispersion at all, the variance will be zero (every value of X will be the same and equal to μ, so the top line in the equation above will be zero). The variance will increase as the dispersion of the values about the mean increases.

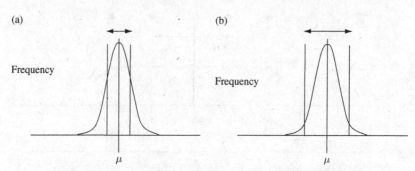

Figure 7.3 Illustration of the proportions of the values in a normally distributed population. (a) 68.27% of values are within ±1 standard deviation from the mean and (b) 95% of values are within ±1.96 standard deviations from the mean. These percentages correspond to the area of the distribution enclosed by the two vertical lines.

7.3.3   The standard deviation of a population

The importance of the variance is apparent when you obtain the standard deviation, which is symbolized for a population by σ and is just the square root of the variance. For example, if the variance is 64, the standard deviation is 8.

The standard deviation is important because the mean of a normally distributed population, plus or minus one standard deviation, includes 68.27% of the values within that population.

Even more importantly, 95% of the values in the population will be within ±1.96 standard deviations of the mean. This is especially useful because the remaining 5% of values will be outside this range and therefore further away from the mean (Figure 7.3). Remember from Chapter 6 that 5% is the commonly used significance level.

These two statistics are all you need to describe the location and shape of a normal distribution and can also be used to determine the proportion of the population that is less than or more than a particular value (Box 7.1).

7.3.4   The Z statistic

The proportions of the normal distribution described in the previous section can be expressed in a different and more workable way. For a normal distribution, the difference between any value and the mean, divided by the standard deviation, gives a ratio called the **Z statistic** that is also normally

---

> **Box 7.1** Use of the standard normal distribution
>
> For a normally distributed population of plagioclase phenocrysts with a mean length of 170 μm and a standard deviation of 10 μm, 95% of these crystals will have lengths in the range from 170 ± (1.96 × 10) μm (which is 150.4 to 189.6 μm). You only have a 5% chance of finding a phenocryst that is either longer than 189.6 μm or shorter than 150.4 μm.

distributed, with a mean of zero and a standard deviation of 1.00. This is called the **standard normal distribution**:

$$Z = \frac{X_i - \mu}{\sigma} \tag{7.3}$$

Consequently, the value of the Z statistic specifies the number of standard deviations it is from the mean. In the example in Box 7.1, a value of 189.6 μm is $\frac{189\cdot6 - 170}{10} = 1.96$ standard deviations away from the mean.

In contrast, a value of 175 μm is $\frac{175 - 170}{10} = 0.5$ standard deviations away from the mean.

When this ratio is greater than +1.96 or less than −1.96, the probability of obtaining that value of X is less than 5%. The Z statistic will be discussed again later in this chapter.

7.4 . Samples and populations

The equations for the mean, variance and standard deviation given above apply to a **population** – the case where you have obtained data for every case or individual that is present. For a population the values of μ, σ^2 and σ are called **parameters** or **population statistics** and are true values (assuming no mistakes in measurement or calculation). Of course in geological situations we rarely have a true population, so μ and σ are not known and must be estimated.

When you take a **sample** from a population and calculate the sample mean, sample variance and sample standard deviation, these are **true values for that sample** but are only **estimates** of μ, σ^2 and σ. Consequently, they are given different symbols (the Roman $\overline{X}$, s^2 and s respectively) and are called **sample statistics**. But remember – because these statistics are only estimates, they may not be accurate measures of the true population statistics.

7.4.1 The sample mean

First, the procedure for calculating a sample mean is the same as for the population mean, except (as mentioned above) the sample mean is symbolized by $\overline{X}$ because it is only an estimate of μ.

The sample mean is:

$$\overline{X} = \frac{\sum\limits_{i=1}^{n} X_i}{n} \tag{7.4}$$

(Note that the lower case n is used to indicate the sample size, compared to the capital N used to indicate the population size in Equation (7.1).)

7.4.2 The sample variance

When you calculate the sample variance, this estimate of σ^2 is also likely to be subject to error. Small sample sizes introduce a consistent bias, but this can be compensated for by a modification to Equation (7.2). For a population, the variance is:

$$\sigma^2 = \frac{\sum\limits_{i=1}^{N} (X_i - \mu)^2}{N} \tag{7.5 copied from 7.2}$$

In contrast, the sample variance is estimated using the following formula:

$$s^2 = \frac{\sum\limits_{i=1}^{n} (X_i - \overline{X})^2}{n - 1} \tag{7.6}$$

Note that the sum of squares is divided by $n - 1$, but you would expect it to be divided by n. This is to reduce a bias caused by the small sample size, and it is easily explained by an example. Imagine you wanted to estimate the population variance of the height of all adult *Stegosaurus* fossils from the sample of the 80 that have been found. This small sample is unlikely to include a sufficient proportion of animals that are in either the upper or lower extremes of height within that population (the really short and really tall animals), because there are relatively few of them. They will, nevertheless, make a big contribution to the population variance because they are so far from the mean (the value of $(X_i - \mu)^2$ will be a large quantity for every one of those individuals). So the sample variance will tend to underestimate the population variance and needs to be corrected.

To illustrate this, we ask our students to look around the lecture room and ask themselves "Are there any extremely tall or short people present?" (The answer so far has been "No." One day, depending on who shows up to our classes, we may have to choose a different variable.) To make s^2 the best possible estimate of σ^2, you need to divide the sum of squares by $n - 1$, not n. This correction will make the sample variance (and sample standard deviation) larger.

Note that this correction will have a considerable effect when n is small (imagine dividing by 3 instead of 4) but less effect as sample size increases (imagine dividing by 999 instead of 1000). Smaller corrections are needed as sample size increases because larger samples are more likely to include individuals from the extremes of the population you are sampling.

Here you may be thinking "Why don't I have to correct the mean in this way as well?" You do not because you are equally likely to miss out on sampling both the positive and negative extremes in the population.

7.5   Your sample mean may not be an accurate estimate of the population mean

A sample mean $(\overline{X})$ may, or may not, be an accurate estimation of the true population mean μ. Estimates from small samples are especially likely to be inaccurate, simply by chance.

To illustrate this, if you take a lot of samples of a certain size (n) at random from a population and calculate the **mean** of each sample, they are unlikely to all be the same. Instead the sample means will be dispersed around the population mean μ.

Statisticians have shown that the distribution of these sample means is also normal **with its own mean (which is also a good estimate of μ), variance** and **standard deviation.**

The standard deviation of the distribution of sample means is an extremely important statistic. It is called **the standard error of the mean,** or the **standard error,** often abbreviated as **SEM** (not to be confused with a scanning electron microscope with the same acronym) or **SE,** and given the symbol $\sigma_{\overline{X}}$ to distinguish it from the sample standard deviation (s) and

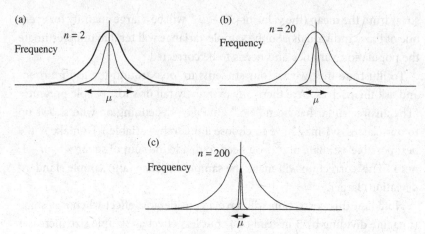

Figure 7.4 The effect of sample size on the precision and accuracy of values of $\overline{X}$ as estimates of μ. The heavy line shows the distribution of a population with parametric mean μ. The lighter line shows the distribution of the means of 200 independent samples, each of which has a sample size of (a) 2, (b) 20 and (c) 200. Note that the distribution of the means is normal with a mean of μ and that the expected range of the sample means decreases as sample size increases. The double-headed arrow shows the range within which 95% of the sample means are expected to occur.

the population standard deviation (σ). Importantly, **as sample size increases the standard error of the mean decreases** and therefore the accuracy of any single estimate of the population mean is likely to improve. This is shown in Figure 7.4.

It is useful to know how precise your estimate ($\overline{X}$) of μ is likely to be for a certain sample size. When you take a lot of samples, each of size n, from a population whose parametric statistics are known (as illustrated in Figure 7.4) the **standard error of the mean** can be estimated by dividing the standard deviation of the population by the square root of the sample size (n):

$$\text{SEM} = \sigma_{\overline{X}} = \frac{\sigma}{\sqrt{n}} \tag{7.7}$$

A numerical example is given in Table 7.1, which clearly illustrates that the means of larger samples are likely to be relatively close to the population mean.

The standard error of the mean is important because it can be used to calculate the range within which a particular percentage of the sample means

Table 7.1 A numerical example of the effect of sample size on the accuracy and precision of values of $\overline{X}$ obtained by taking random samples of size 2, 20 and 200 from a population with a known variance of 600. As sample size increases the values of the sample means become much closer to the population mean. Precision improves and therefore the sample means will tend to be more accurate estimates of μ.

Population parameters				
Variance σ^2	σ	Sample size (n)	$\sqrt{n}$	Standard error of the mean ($\frac{\sigma}{\sqrt{n}}$)
600	24.49	2	1.41	17.32
600	24.49	20	4.47	5.48
600	24.49	200	14.14	1.73

will occur. Because the sample means are normally distributed with a mean of μ, then $\mu \pm 1$ SEM will include 68.27% of the sample means and $\mu \pm 1.96$ SEM will include 95% of the sample means.

This can also be expressed as a ratio. The difference between any sample mean $\overline{X}$ and the population mean μ, divided by the standard error of the mean:

$$\frac{\overline{X} - \mu}{\sigma_{\overline{X}}} \tag{7.8}$$

will give the Z statistic already discussed in Section 7.3.4, with a mean of zero and a standard deviation of 1.00. As the difference between $\overline{X}$ and μ increases the value of Z will become increasingly positive (if $\overline{X}$ is greater than μ) or increasingly negative (if $\overline{X}$ is less than μ). Once Z is less than -1.96, or greater than $+1.96$, the probability of getting that difference between the sample mean and the known population mean μ is less than 5% (Figure 7.5).

This formula can be used to test hypotheses about the means of samples when population parameters are known. Box 7.2 gives a worked example.

7.6 What do you do when you only have data from one sample?

As shown above, the standard error of the mean is very important for hypothesis testing because it can be used to predict the range around μ within which 95% of means of a certain sample size will occur.

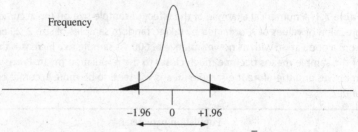

Figure 7.5 Distribution of the Z statistic (the ratio of $\frac{\overline{X}-\mu}{\text{SEM}}$ obtained by taking the means of a large number of small samples from a normal distribution). By chance 95% of the sample means will be within the range −1.96 to +1.96 (the unshaded area), with the remaining 5% outside this range (the two symmetrical shaded areas).

Box 7.2 Use of the Z statistic

The known population value of μ is 100 and σ is 36. You take a sample of 16 individuals and obtain a sample mean of 81. What is the probability that this sample is from the population?

$\mu = 100$, $\sigma = 36$, $n = 16$, so the $\sqrt{n} = 4$, and the SEM $= \sigma/\sqrt{n} = 36/4 = 9$

Therefore the value of:

$$\frac{\overline{X} - \mu}{\text{SEM}} \quad \text{is} \quad \frac{81 - 100}{9} = -2.11$$

The ratio is outside the range of ±1.96, so the probability that the sample mean has come from a population with a mean of μ is less than 0.05. Thus, the sample mean is significantly different to the population mean.

Unfortunately, a researcher usually does not know the true values of the population parameters μ and σ **because they only have a sample,** and statistical decisions have to be made from the limited information provided by that sample. Here too, knowing the standard error of the mean would be extremely helpful!

If you only have data from a single sample, you can calculate the sample mean ($\overline{X}$), the sample variance (s^2) and sample standard deviation (s). These are your **best estimates** of the population statistics μ, σ^2 and σ. Therefore, you can use s to **estimate** the standard error of the mean by substituting s for

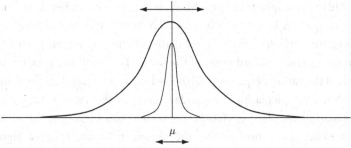

Figure 7.6 If you only have one sample, you can calculate the standard deviation, *s*, which is your only estimate of the population standard deviation σ. You can estimate the standard error of the mean of the population by dividing the sample standard deviation by the square root of the sample size (Equation (7.9)). The lower, shorter double-headed arrow shows the range within which 95% of the means of all samples of size *n* taken from a population with a hypothetical mean of μ would be expected to occur.

σ in Equation (7.7). This is also called the standard error of the mean and is abbreviated as "SEM" as noted earlier:

$$\text{SEM} = s_{\overline{X}} = \frac{s}{\sqrt{n}} \tag{7.9}$$

where *s* is the sample standard deviation and *n* is the sample size. Note from Equation (7.9) that the sample SEM estimated in this way has a different symbol to the SEM estimated from the population statistics ($s_{\overline{X}}$ instead of $\sigma_{\overline{X}}$).

What does this give you? The estimate of the standard error of the mean, made from your sample, can be used to predict the range around any hypothetical value of μ within which 95% of the means of all samples of size *n* taken from that population will occur. This is shown in Figure 7.6.

Therefore, in terms of making a decision about whether your sample mean differs significantly from an expected value of μ, the formula:

$$\frac{\overline{X} - \mu_{\text{expected}}}{\text{SEM}} \tag{7.10}$$

corresponds to Equation (7.8), but with $s_{\overline{X}}$ used instead of $\sigma_{\overline{X}}$ as the SEM. Here it seems logical that when this ratio is < −1.96 or > + 1.96, the difference between the sample mean and the expected value would be considered statistically significant at the 5% level.

This is an appropriate procedure, **but a correction is needed**, especially for samples of less than 100, which are very prone to sampling error and therefore likely to give poor estimates of the population mean, standard deviation and standard error of the mean. For small samples, the distribution of the ratio in Equation (7.10) is **wider and flatter** than the distribution obtained by calculating the standard error of the mean from the (known) population standard deviation. As sample size increases, the distribution gets closer and closer to the one shown in Figure 7.5 (see Figure 7.7). Therefore Equation (7.10) is appropriate, but for small samples the range within which 95% of the values of all means will occur is **wider** (e.g. for a sample size of 4 the adjusted range within which 95% of values would be expected is from −3.182 to +3.182). **Using this correction, you can test hypotheses about your sample mean $\overline{X}$ without knowing the population statistics.**

The shape of this wider and flatter distribution of the expected ratio for small samples was established by W. S. Gossett who published his work under the pseudonym of "Student" (see Student, 1908). Consequently the distribution is often called the "Student" distribution or "Student's *t*" distribution. Two examples of the distribution of *t* are shown in Figure 7.7 and Table 7.2. As sample size increases, the *t* statistic for an α of 0.05 decreases and becomes closer and closer to 1.96, which is the value for a sample of infinite size and also for the *Z* statistic.

7.7   Why are the statistics that describe the normal distribution so important?

Sample statistics like the mean, variance, standard deviation, and **especially the standard error of the mean** are estimates of population statistics that can be used to predict the range within which 95% of the means of a particular sample size will occur. Knowing this, you can use a parametric test to estimate the probability that a sample mean is the same as an expected value, or the probability that the means of two samples are from the same population. These tests will be described in Chapter 8.

Here you might be thinking "These statistical methods have the potential to be very prone to error! My sample mean may be an inaccurate estimate of μ and then I'm using the sample standard deviation (*s*) to infer the standard error of the mean." This is true and unavoidable when you extrapolate from

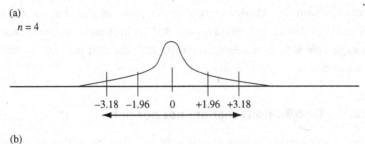

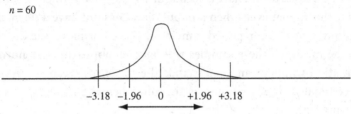

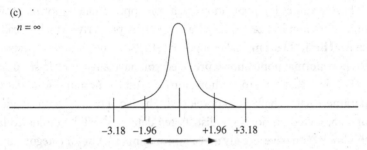

Figure 7.7 Illustration of the distribution of the t statistic obtained when the sample statistic s is used as an estimate of σ (a) for $n = 4$, (b) for $n = 60$ and (c) $n = \infty$.

Table 7.2 The range of the 95% confidence interval for the t statistic in relation to sample size. (a) $n = 4$, (b) $n = 60$, (c) $n = 200$ (d) $n = 1000$ and (e) $n = \infty$. Note that the 95% confidence interval decreases as the sample size increases, and the value of t for a sample of infinite size is the same as the Z statistic. Values of t for finite degrees of freedom were calculated using the equations given by Zelen and Severo (1964).

	Formula	Statistic	Sample size	95% confidence interval
(a)	$\frac{\overline{X}-\mu}{s_{\overline{X}}}$	t	4	±3.182
(b)	$\frac{\overline{X}-\mu}{s_{\overline{X}}}$	t	60	±2.001
(c)	$\frac{\overline{X}-\mu}{s_{\overline{X}}}$	t	200	±1.972
(d)	$\frac{\overline{X}-\mu}{s_{\overline{X}}}$	t	1000	±1.962
(e)	$\frac{\overline{X}-\mu}{s_{\overline{X}}}$	t	∞	±1.96

only one sample, but the corrections described in this chapter and knowledge of how the sample mean is likely to become a more accurate estimate of μ as sample size increases, help ensure that the best possible estimates are obtained.

7.8   Distributions that are not normal

Some variables do not have a normal distribution. Nevertheless, statisticians have shown that **even when a population does not have a normal distribution**, if you take repeated samples of size 25 or more, **the distribution of the means of these samples will have an approximately normal distribution** with a mean μ and standard error of the mean $\sigma/\sqrt{n}$ (that can be estimated by $s/\sqrt{n}$), just as they do when the population is normal (Figure 7.8).

Furthermore, for populations that are approximately normal, this even holds for sample sizes as small as five. This property, which is called the **central limit theorem**, makes it possible to use some parametric tests on data for non-normal populations, provided you have a reasonable-sized sample.

For data that are grossly non-normal, and for nominal scale data, **non-parametric** tests have been developed. These can be used with a wide range of data, including normally distributed data, and will be discussed later in this book. You have already met a non-parametric test for categorical data in Chapter 6 when the chi-square test was used to compare the observed and expected proportions in two categories.

7.9   Other distributions

Not all data are normally distributed. Sometimes a frequency distribution may resemble a normal distribution and be symmetrical but much flatter (Figure 7.9(b)). This type of distribution is **platykurtic**. In contrast, a distribution that resembles a normal distribution but has too many values around the mean and in the tails is **leptokurtic** (Figure 7.9(c))

If a distribution is similar to a normal one but not symmetrical in that one of its tails extends further than the other, it is **skewed**. If the upper tail is longer the distribution has a **positive skew** (Figure 7.9(d)) and if the lower tail is longer it has a **negative skew**. Other distributions include the binomial distribution and the Poisson distribution.

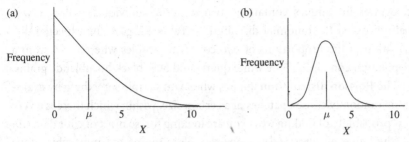

Figure 7.8 An example of the central limit theorem. Even if a population does not have a normal distribution, samples of size 25 (or greater) from that population will have an approximately normal distribution with mean μ and standard error of $\sigma/\sqrt{n}$ (that can be estimated from a sample by $s/\sqrt{n}$). (a) Distribution of a population that is not normal, with mean μ and standard deviation σ. (b) The distribution of 200 samples, each of $n = 25$ taken at random from the population shown in (a), is approximately normal with a mean of μ and standard error of $\sigma/\sqrt{n}$.

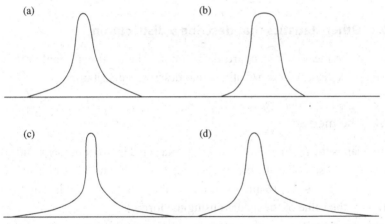

Figure 7.9 Distributions that are similar to the normal distribution. (a) a normal distribution, (b) a platykurtic distribution, (c) a leptokurtic distribution, (d) positive skew.

The binomial distribution has already been mentioned in Chapter 6. If a population can be partitioned into two categories (e.g. white quartz and black hornblende grains in beach sand) then the probability of sampling **either** category is 1.0 and the probability of sampling a particular category will be its proportion in the population (e.g. 0.5 for a population where half the grains are black and half are white). The proportions of each of the two categories in samples containing two or more individuals will follow a pattern called the binomial distribution. Table 6.2 gave the expected distribution of the proportions of two colors in samples where $n = 6$ from a population containing 50% white quartz and 50% black hornblende grains.

The Poisson distribution applies when you sample something by examining randomly chosen patches of a certain size, within which there is a very low probability of finding what you are looking for, so most of your data will be the value of zero. Here is an example. One of the most ubiquitous geological processes is volcanism which is why basalts are one of the most common rock types on the Earth's surface. Despite this, active volcanoes are quite rare. Therefore, if you were to subdivide the surface of the Earth into a large number of squares, each of 10 000 km^2 in area, most will not contain any active volcanoes. Occasionally, however, a square will contain one, two or even more rarely three or more. This will generate a Poisson distribution where most values are zero, a few are "1" and even fewer are "2" and "3" etc.

7.10   Other statistics that describe a distribution

Although the mean and standard deviation are the most commonly used descriptive statistics, there are others that describe a distribution.

7.10.1   The median

The median is the middle value of a set of data listed in order of magnitude. For example, a sample with the values 1, 6, 3, 9, 4, 16 and 11 is ranked in order as 1, 3, 4, 6, 9, 11, 16 and the middle value is 6. You can calculate the location of the value of the median using the formula:

$$M = X_{(n+1)/2} \tag{7.11}$$

This means "The median is the value of X whose numbered position in an ordered sequence corresponds to the sample size plus one, and then divided by two." For the sample of seven listed above, the median is the fourth value, X_4, which is 6. For even-sized samples, the median will lie between two values (e.g. $X_{5.5}$), in which case it is the average of the values below (X_5) and above it (X_6). The procedure becomes more complex when there are tied values, but most statistical packages will readily calculate the median of a set of data.

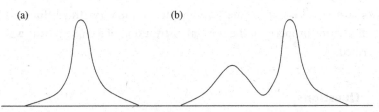

Figure 7.10 (a) A unimodal distribution. (b) A bimodal distribution.

7.10.2 The mode

The mode is defined as the most frequently occurring value in a set of data, so a normal distribution has only one mode. Sometimes, however, a distribution may have two or more clearly separated peaks in which case it is bimodal or multimodal respectively (Figure 7.10).

7.10.3 The range

The range is the difference between the largest and smallest value in a sample or population. The range of the set of data in Section 7.10.1 is $16-1 = 15$.

7.11 Conclusion

The mean and the standard deviation are the only statistics needed to describe the shape of a normal distribution. The sample statistics $\overline{X}$ and s provide estimates of the population statistics μ and σ. Importantly, the distribution of the means of samples from a normal population is also normal, with a mean of μ and a standard error of $\sigma/\sqrt{n}$ that can be estimated from a sample of two or more by $s/\sqrt{n}$. This allows you to use the properties of the normal distribution to predict the range around $\overline{X}$ (your best and only estimate of μ) within which 95% (or 99% or 99.9% if required) of the means of all samples of size n taken from that population will occur.

Even more importantly, when the population of the variable you have measured is not normally distributed, the distribution of the means of samples of about 25 or more will be approximately normal, with a mean of μ and a standard error of $\sigma/\sqrt{n}$. This also provides a way of predicting the range of values within which there is a 95% probability that any sample mean of size n will occur. In the next chapter, some very straightforward tests that use this property of the normal distribution of sample means will be described.

7.12   Questions

(1) It is known that a **population** of the fossil snail *Calcarus porosis* in Bentley County, South Dakota, has a mean shell length of 100 mm and a standard deviation of 10 mm. A paleontologist measured one fossil snail from this population and found it had a shell length of 78 mm. The paleontologist said "This is an impossible result." Please comment on what they said, including whether you agree or disagree, and why.

(2) Why does the variance calculated from a sample have to be corrected to give a realistic indication of the variance of the population from which it has been taken?

8 | Normal distributions: tests for comparing the means of one and two samples

8.1   Introduction

Although sample statistics such as $\overline{X}$ and s are only estimates of population statistics, it is still possible to use these to make statistical decisions. First, as sample size increases, sample statistics are likely to become increasingly accurate estimates of population statistics. Second, as described in Chapter 7, the distribution of the means of samples of a particular size (n) taken from a normal population with population statistics of μ and σ will also be normal, with a mean of μ and a standard error of the mean of $\sigma/\sqrt{n}$ that can be estimated from a sample by $s/\sqrt{n}$. Even more usefully, provided you have a sample size of about 25 or more, these properties of the distribution of sample means apply, even when the population they have been taken from is not normal, provided it is not grossly so (e.g. a bimodal distribution). Therefore, you can often use a parametric test to make decisions about sample means even when the population you have sampled is not normally distributed.

In this chapter, these concepts are used to describe how some parametric tests for comparing the means of one and two samples actually work. The first test is for comparing a single-sample mean to a known population mean. The second is for comparing a single-sample mean to a hypothesized value. These are followed by tests for comparing the means of two samples.

8.2   The 95% confidence interval and 95% confidence limits

In Chapter 7, we discussed how 95% of the means of a sample size n, taken from a population with known μ and σ, would be expected to occur **within** the range of $\mu \pm 1.96 \times$ SEM. This range is called the **95% confidence interval**, and the actual numbers that show the limits of that range ($\mu \pm 1.96 \times$ SEM) are called the **95% confidence limits**.

If you only have data for one sample of size n, then the sample standard deviation s is your best estimate of σ, and it can be used with the appropriate t statistic to calculate the 95% confidence interval for an expected or hypothesized value of μ. You have to use the formula $\mu_{expected} \pm t \times$ SEM because the population statistics are not known. This formula will give a wider confidence interval than if population statistics are known because the value of t for a finite sample size is always greater than 1.96, especially for small samples (Chapter 7).

8.3   Using the Z statistic to compare a sample mean and population mean when population statistics are known

This test uses the Z statistic to give the probability that a sample mean has been taken from a population with a known mean and standard deviation. From the population statistics μ and σ, you can calculate the expected standard error of the mean $(\sigma/\sqrt{n})$ for a sample of size of n and therefore the 95% confidence interval (Figure 8.1), which is the range within $\mu \pm 1.96 \times$ SEM. If your sample mean, $\overline{X}$, occurs within this range, then the probability that it has come from the population with a mean of μ is 0.05 or greater. **So, the mean of the population from which the sample has been taken is not significantly different to the known population mean.** If, however, your sample mean occurs outside the confidence interval, the probability that it has been taken from the population of mean μ is less than 0.05. **So, the mean of the population from which the sample has been taken is significantly different to the known population mean μ.**

This is a very straightforward test (Figure 8.1). If you decide on a probability level other than 0.05, you simply need to use a different value than 1.96 (e.g. for the 99% confidence interval you would use 2.576).

Although you could calculate the 95% confidence limits every time you made this type of comparison, it is far easier to calculate the ratio $Z = \frac{\overline{X}-\mu}{\text{SEM}}$ as described in Section 7.3.4. All this formula does is divide the distance between the sample mean and the known population mean by the standard error. If the value of Z is < -1.96 or $> +1.96$, the mean of the population from which the sample has been taken is considered significantly different to the known population mean, assuming an $\alpha = 0.05$.

Here you may be wondering if a population mean could ever be known, apart from small populations where every individual has been considered.

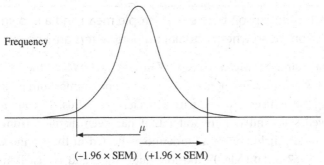

Figure 8.1 The 95% confidence interval, obtained by taking the means of a large number of small samples from a normally distributed population with known statistics is indicated by the horizontal distance enclosed within $\mu \pm$ 1.96 SEM. The remaining 5% of sample means are expected to be further away from μ. Therefore, a sample mean that lies **inside** the 95% confidence interval will be considered to have come from the population with a mean of μ, while a sample mean that lies **outside** the 95% confidence interval will be considered to have come from a population with a mean significantly different to μ, assuming an $\alpha = 0.05$.

Sometimes, however, researchers have so many data for a particular variable that they consider the sample statistics indicate the true values of population statistics. For example, many important physical parameters such as seismic velocities of key rock types, rare earth element abundances in chondrites (a primitive type of meteorite), and the isotopic composition of Vienna Standard Mean Ocean Water (VSMOW) have been measured repeatedly, hundreds of thousands of times. These sample sizes are so large that they can be considered to give extremely accurate estimates of the population statistics. Remember that as sample size increases, $\overline{X}$ becomes closer and closer to the true population mean and the correction of $n - 1$ used to calculate the standard deviation also becomes less and less important. There is an example of the comparison between a sample mean and a "known" population mean in Box 8.1.

8.4 Comparing a sample mean to an expected value when population statistics are not known

The single-sample t test compares a single-sample mean to an expected value of the population mean. When population statistics are not known, the sample standard deviation s is your best and only estimate of σ for the population from which it has been taken. You can still use the 95%

Box 8.1 Comparison between a sample mean and a known population mean where population parameters are known

Vienna Standard Mean Ocean Water (VSMOW) is the standard against which measurements of oxygen isotopes in most other oxygen-bearing substances are compared, usually as ratios. It contains no dissolved salts and is pure water that has been distilled from deep ocean water, including small amounts collected in the Pacific Ocean in July 1967 at latitude 0° and longitude 180°, and is distributed by the US National Institute of Standards and Technology on behalf of the International Atomic Energy Agency, Vienna, Austria (thus the name). The population mean for the ratio of $^{18}O/^{16}O$ in VSMOW is 2005.20×10^6, with a standard deviation of 0.45×10^6. (There are no units given here because it is a ratio.) These statistics are from a very large sample of measurements and are therefore considered to be the population statistics μ and σ.

On a recent traverse of the same area of the Pacific, also in the month of July, you have collected 10 water samples. The data are shown below. What is the probability that your sample mean $\overline{X}$ is the same as that of the VSMOW population?

Your measured $^{18}O/^{16}O$ ratios are: 2005.23, 2006.13, 2007.66, 2006.98, 2003.24, 2004.45, 2005.57, 2003.34, 2005.6 and 2005.01 (all $\times 10^6$).

The population statistics for VSMOW are $\mu = 2005.20 \times 10^6$ and $\sigma = 0.45 \times 10^6$. Because all values are to the power of 10^6 this has been left out of the following calculation to make it easier to follow.

The sample size $n = 10$

The sample mean $\overline{X} = 2005.32$

The standard error of the mean $= \frac{\sigma}{\sqrt{n}} = \frac{0.45}{\sqrt{10}} = 0.142$

Therefore, $1.96 \times SEM = 1.96 \times (0.142) = 0.28$, so the 95% confidence interval for the means of samples of $n = 10$ is 2005.20 ± 0.28, which is from 2004.92 up to 2005.48. Because the mean $^{18}O/^{16}O$ ratio of your ten replicates (2005.32) lies within the range in which 95% of means with $n = 10$ would be expected to occur, the mean of the population from which the samples have been taken does not differ significantly from the VSMOW population.

Expressed as a formula:

$$Z = \frac{\overline{X} - \mu}{SEM} = \frac{2005.32 - 2005.20}{0.142} = \frac{0.12}{0.142} = 0.86$$

Here too, because the Z value lies within the range of ± 1.96, the mean of the population from which the sample has been taken does not differ significantly from the mean of the VSMOW population.

confidence interval of the mean, estimated from the sample standard deviation, and the t statistic described in Chapter 7 to predict the range around an expected value of μ within which 95% of the means of samples of size n taken from that population will occur. Here too, once the sample mean lies outside the 95% confidence interval, the probability of it being from a population with a mean of $\mu_{expected}$ is less than 0.05 (Figure 8.2).

Expressed as a formula, as soon as the ratio of $t = \frac{\overline{X} - \mu_{expected}}{SEM}$ is less than the critical 5% value of $-t$ or greater than $+t$, then the sample mean is considered to have come from a population with a mean significantly different to $\mu_{expected}$.

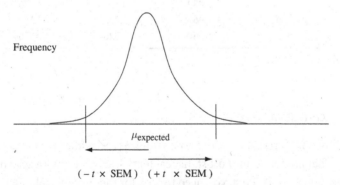

$$(-t \times SEM) \quad (+t \times SEM)$$

Figure 8.2 The 95% confidence interval, estimated from one sample of size n by using the t statistic, is indicated by the horizontal distance enclosed within $\mu_{expected} \pm t \times SEM$. Therefore, 5% of the means of sample size n from the population would be expected to lie outside this range, and if $\overline{X}$ lies inside the confidence interval, it will be considered to have come from a population with a mean the same as $\mu_{expected}$. If it lies outside the confidence interval it will be considered to have come from a population with a significantly different mean, assuming an $\alpha = 0.05$.

Table 8.1 Critical values of the distribution of t. The column on the far left gives the number of degrees of freedom (v). The remaining columns give the critical value of t. For example, the third column, shown in bold and headed $\alpha(2) = 0.05$, gives the 5% critical values. Note that the 5% probability value of t for a sample of infinite size (the last row) is 1.96 and thus equal to the 5% probability value for the Z distribution. Finite critical values were calculated using the methods given by Zelen and Severo (1964). A more extensive table is given in Appendix A.

Degrees of freedom v	$\alpha(2) = 0.10$ or $\alpha(1) = 0.05$	$\alpha(2) = 0.05$ or $\alpha(1) = 0.025$	$\alpha(2) = 0.025$ or $\alpha(1) = 0.01$	$\alpha(2) = 0.01$ or $\alpha(1) = 0.005$
1	6.314	**12.706**	31.821	63.657
2	2.920	**4.303**	6.965	9.925
3	2.353	**3.182**	4.541	5.841
4	2.132	**2.776**	3.747	4.604
5	2.015	**2.571**	3.365	4.032
6	1.934	**2.447**	3.143	3.707
7	1.895	**2.365**	2.998	3.499
8	1.860	**2.306**	2.896	3.355
9	1.833	**2.262**	2.821	3.250
10	1.812	**2.228**	2.764	3.169
15	1.753	**2.131**	2.602	2.947
30	1.697	**2.042**	2.457	2.750
50	1.676	**2.009**	2.403	2.678
100	1.660	**1.984**	2.364	2.626
1000	1.646	**1.962**	2.330	2.581
∞	1.645	**1.960**	2.326	2.576

8.4.1   Degrees of freedom and looking up the appropriate critical value of t

The appropriate critical value of t for a sample is easily found in tables of this statistic that are found in most statistical texts. Table 8.1 gives a selection of values as an example. First, you need to look for the chosen probability level along the top line labelled as $\alpha(2)$. (There will shortly be an explanation for the column heading $\alpha(1)$.) Here, we are using $\alpha = 0.05$ and the column giving these critical values is shown in bold.

The column on the left gives the number of **degrees of freedom,** which needs explanation. If you have a sample of size n and the mean of this sample is a specified value, then **all of the data within the sample except**

one are free to be any number at all, but the final one is fixed because the sum of the data in the sample, divided by n, must equal the mean.

Here is an example. If you have a specified sample mean of 4.25 and $n = 2$, then the first value in the sample is free to be any value at all, but the second must be one that gives a mean of 4.25, so it is a fixed number. Thus, the number of degrees of freedom for a sample of $n = 2$ is 1. For $n = 100$ and a specified mean (e.g. 4.25), 99 of the values are free to vary, but the final value is also determined by the requirement for the mean to be 4.25, so the number of degrees of freedom is 99.

The number of degrees of freedom determines the critical value of the t statistic. For a single-sample t test, if your sample size is n, then you need to use the t value that has $n - 1$ degrees of freedom. Therefore, for a sample size of 10, the degrees of freedom are 9 and the critical value of the t statistic for an $\alpha = 0.05$ is 2.262 (Table 8.1). If your calculated value of t is less than -2.262 or more than $+2.262$, then the expected probability of that outcome is < 0.05. From now on, the appropriate t value will have a subscript to show the degrees of freedom (e.g. t_7 indicates 7 degrees of freedom).

8.4.2   One-tailed and two-tailed tests

All of the alternate hypotheses dealt with so far in this chapter do not specify anything other than "The mean of the population from which the sample has been drawn is different to an expected value" or "The two samples are from populations with different means." Therefore, these are **two-tailed hypotheses** because nothing is specified about the **direction** of the difference. The null hypothesis could be rejected by a difference in either a positive or negative direction.

Sometimes, however, you may have an alternate hypothesis that specifies a direction. For example, "The mean of the population from which the sample has been taken is **greater** than an expected value" or "The mean of the population from which sample A has been taken is **less** than the mean of the population from which sample B has been taken." These are called **one-tailed hypotheses**.

If you have an alternate hypothesis that is directional, the null hypothesis will not just be one of no difference. For example, if the alternate hypothesis states that the mean of the population from which the sample has been taken will be **less** than an expected value, then the null should state, "The

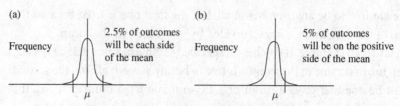

Figure 8.3 The distribution of the 5% of most extreme outcomes under a two-tailed hypothesis and a one-tailed hypothesis specifying that the expected value of the mean is larger than μ. (a) The rejection regions for a two-tailed hypothesis are on both the positive and negative sides of the true population mean. (b) The rejection region for a one-tailed hypothesis occurs only on one side of the true population mean. Here it is on the right side because the hypothesis specifies that the sample mean is taken from a population with a larger mean than μ.

mean of the population from which the sample has been taken will be no different to, or **more**, than the expected value."

You need to be cautious, however, because a directional hypothesis will affect the location of the region where the most extreme 5% of outcomes will occur. Here is an example using a single-sample test where the true population mean is known. For any two-tailed hypothesis the 5% rejection region is split equally into two areas of 2.5% on the negative and positive side of μ (Figure 8.3(a)).

If, however, the hypothesis specifies that your sample is from a population with a mean that is expected to be only greater (or only less) than the true value, then in each case the most extreme 5% of possible outcomes that you would be interested in are restricted to **one side** or **one tail** of the distribution (Figure 8.3(b)).

Therefore, if you have a one-tailed hypothesis you need to do two things to make sure you make an appropriate decision.

First, you need to examine your results to see if the difference is in the direction expected under the alternate hypothesis. **If it is not then the value of the t statistic is irrelevant** – the null hypothesis will stand and the alternate hypothesis will be rejected (Figure 8.4).

Second, if the difference is in the appropriate direction, then you need to choose an appropriate critical value to ensure that 5% of outcomes are concentrated in one tail of the expected distribution. This is easy. For the Z or t statistics, the critical probability of 5% is not appropriate for a one-tailed test because it only specifies the region where 2.5% of the values will

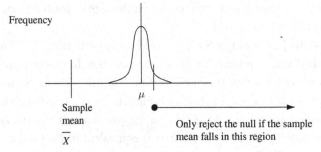

Figure 8.4 An example of the rejection region for a one-tailed test. If the alternate hypothesis states that the sample mean will be more than μ, then the null hypothesis is retained unless the sample mean lies in the region to the right where the most extreme 5% of values would be expected to occur.

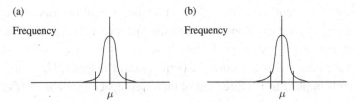

Figure 8.5 (a) A two-tailed test using the 5% probability level will have a rejection region of 2.5% on both the positive and negative sides of the known population mean. The positive and negative of the critical value will define the region where the null hypothesis is rejected. (b) A one-tailed test using the 5% probability level will have a rejection region of 5% on only one side of the population mean. Therefore the 5% critical value will correspond to the value for a 10% two-tailed test, except that it will only be either the positive or negative of the critical value, depending on the direction of the alternate hypothesis.

occur in each tail. So to get the critical 5% value for a one-tailed test, you would need to use the 10% critical value for a two-tailed test. This is why the column headed $\alpha(2) = 0.10$ in Table 8.1 also includes the heading $\alpha(1) = 0.05$, and you would need to use the critical values in this column if you were doing a one-tailed test.

It is important to specify your null and alternate hypotheses, and therefore decide whether a one- or two-tailed test is appropriate, **before** you do an experiment, because the critical values are different. For example, for an $\alpha = 0.05$, the two-tailed critical value for t_{10} is ±2.228 (Table 8.1), but if the test were one-tailed, the critical value would be **either** +1.812 or −1.812. So a

t value of 2.0 in the correct direction would be significant for a one-tailed test but not for a two-tailed test.

Many statistical packages only give the calculated value of t (not the critical value) and its probability for a two-tailed test. In this case, however, it is even easier to obtain the one-tailed probability and you do not even need a table of critical values such as Table 8.1. All you have to do is halve the two-tailed probability to get the appropriate one-tailed probability (e.g. a two-tailed probability of $P = 0.08$ is equivalent to $P = 0.04$, provided the difference is in the right direction).

There has been considerable discussion about the appropriateness of one-tailed tests, because the decision to propose a directional hypothesis implies that an outcome in the opposite direction is of **absolutely no interest** to either the researcher or science, but often this is not true. For example, a geoscientist hypothesized that ^{60}Co irradiation would **increase** the opacity of amethyst crystals. They measured the opacity of 10 crystals, irradiated them and then remeasured their opacity. Here, however, if opacity decreased markedly, this outcome (which would be ignored by a one-tailed test only applied in the direction of increased opacity) might be of considerable scientific interest and have industrial application. Therefore, it has been suggested that two-tailed tests should only be applied in the rare circumstances where a one-tailed hypothesis is truly appropriate because there is no interest in the opposite outcome (e.g. evaluation of a new type of fine particle filter in relation to existing products, where you would only be looking for an improvement in performance).

Finally, if your hypothesis is truly one-tailed, it is appropriate to do a one-tailed test. There have, however, been cases of unscrupulous researchers who have obtained a result with a non-significant two-tailed probability (e.g. $P = 0.065$) but have then realized this would be significant if a one-tailed test were applied ($P = 0.0325$) and have subsequently modified their initial hypothesis. This is neither appropriate nor ethical as discussed in Chapter 5.

8.4.3 The application of a single-sample t test

Here is an example where you might use a single-sample t test. The minerals in the vermiculite and smectite groups are the so-called "swelling clays," in which some fraction of the sites between the layers in the structure is filled with cations, leaving the remainder available to be occupied by H_2O

molecules. When vermiculite is heated to about $870\,°C$ the H_2O in the crystal structure expands and is eventually released as steam. The pressure generated by this change of state pushes the layers apart in a process called exfoliation, and can expand the volume by 8–30 times. Vermiculite treated in this way is light and slightly compressible and has long been used for packing insulation and a soil additive.

If you are processing vermiculite you need to monitor the water content (and impurity) of the material very carefully before heating, in order to produce small light pieces. If too little water is present, the vermiculite will only exfoliate slightly, giving dense lumps, but too much water will produce fragments that are very small and powdery. Suppose you know from experience that the desired mean water content for optimal expansion at exfoliation is 7.0 wt% H_2O. A new mine has just opened, and the operators have brought you a sample of nine replicates, collected from widely dispersed parts of their deposit, and offered to sell their product to you for a very reasonable price. You measure the water content of these nine sampling units, and the data are given in Box 8.2.

Box 8.2 Comparison between a sample mean and an expected value when population statistics are not known

The water content of a sample of nine vermiculites taken at random from within the new deposit is 6.1, 5.5, 5.3, 6.8, 7.6, 5.3, 6.9, 6.1 and 5.7 wt% H_2O.

The null hypothesis is that this sample is from a population with a mean water content of 7.0 wt% H_2O.

The alternate hypothesis is that this sample is from a population with a mean water content that is **not** 7.0 wt% H_2O.

The mean of this sample is: 6.14

The standard deviation $s = 0.803$

The standard error of the mean is $\frac{s}{\sqrt{n}} = \frac{0.803}{3} = 0.268$

Therefore $t_8 = \dfrac{\overline{X} - \mu_{expected}}{SEM} = \dfrac{6.14 - 7.0}{0.268} = -3.20.$

Although the mean of the sample (6.14) is close to the desired mean value of 7.0 wt% H_2O, is the difference significant? The calculated value of t_8 is –3.20. The critical value of t_8 for an α of 0.05 is ± 2.306 (Table 8.1).

> Therefore, the probability that the sample mean has been taken from a population with a mean water content of 7.0 wt% H_2O is < 0.05. The vermiculite processor concluded that the mean moisture content of the samples from the new mine was significantly less than that of a population with a mean of 7.0 wt% H_2O and refused the offer of the cheap vermiculite.

Is the sample likely to have come from a population where $\mu = 7.0$ wt% H_2O? The calculations are in Box 8.2 and are straightforward. If you analyze these data using a statistical package, the results will usually include the value of the t statistic and the probability, making it unnecessary to use a table of critical values.

8.5   Comparing the means of two related samples

The paired-sample t test is designed for cases where you have measured the same variable twice on each sampling unit under two different conditions. Here is an example. Coarse-grained rocks such as granites are difficult to analyze chemically because their composition is very heterogeneous. The standard method is to crush a sample to pea-size fragments and pulverize these in a mill (called a shatterbox) to produce a fine homogeneous powder of $< 25 \, \mu m$. You quickly discover that running the shatterbox for more than 60 seconds creates $< 25 \, \mu m$ powders, but these are difficult to handle, intractable to sieve, and messy to clean up. By accident you find that 30 seconds in the shatterbox will give you coarser powders ($< 125 \, \mu m$) that can be sieved without difficulty and clean up easily. If the two grain sizes give the same result when chemically analyzed, you would only have to prepare the coarser one and thereby save a lot of time and effort. Therefore you measure the iron (Fe) content of the **same** 10 granites processed by each method. The results are shown in Table 8.2.

Here the two groups are not independent because the same granites are in each. Nevertheless, you can generate a single independent value for each individual by taking their "$< 25 \, \mu m$" reading away from the "$< 125 \, \mu m$" reading. This will give a single column of differences for the 10 units, which will have its own mean and standard deviation (Table 8.2).

The null hypothesis is that there is no difference between the FeO content of the two grain sizes. Therefore, if the null hypothesis were true, you would

Table 8.2 The Wt% FeO content of ten granites ground to two different grain sizes. The column headed "Difference" gives the Fe content of the < 125 μm fraction minus that of the < 25 μm fraction for each, and the sample statistics are for this column of data.

Granite Number	Wt% FeO content of granites		
	< 25 μm	< 125 μm	Difference
1	13.5	13.6	+0.1
2	14.6	14.6	0.0
3	12.7	12.6	− 0.1
4	15.5	15.7	+0.2
5	11.1	11.1	0.0
6	16.4	16.6	+0.2
7	13.2	13.2	0.0
8	19.3	19.5	+0.2
9	16.7	16.8	+0.1
10	18.4	18.7	+0.3
			$\overline{X} = 0.100$
			$s = 0.1247$
			$n = 10$
			$\mathrm{SEM} = 0.0394$

expect the population of values for the **difference** for each granite to have a mean of zero, and a standard error that can be estimated from the sample of differences by $s/\sqrt{n}$. **This is just another case of a single-sample t test (Section 8.4), but here the expected population mean is zero.** Consequently, all you need to do is calculate the ratio of $\frac{\overline{X}-0}{\mathrm{SEM}}$ and see if this statistic lies within or outside the region where 95% of the means of this sample size would be expected to occur around an expected population mean of zero. This has been done in Box 8.3.

Unfortunately, the result in Box 8.3 tells you that you get a different result with the coarse powder compared to the finer one specified by the standard method. The extra effort needed to create the finer-grained, messy powder appears to be necessary. Although there may be other aspects of experimental design (including the distribution of grain sizes in each pellet, the starting grain size in each split, and the presumption that the starting splits of each granite were identical) that have confounded the results, this result is consistent with the alternate hypothesis.

Box 8.3 A worked example of a paired-sample t test using the data from Table 8.2

$\overline{X} = 0.100$

$s = 0.12472$

$n = 10$

$\text{SEM} = 0.03944$

Therefore $t_9 = \frac{0.10 - 0}{0.03944} = 2 \cdot 5355$

From Table 8.1 the critical value of t_9 is 2.262. Therefore the value of t lies outside the range within which you would expect 95% of t statistics generated by samples of $n = 10$ from a population where $\mu = $ zero, so it was concluded that the mean of the population of the differences in FeO contents was significantly different ($P < 0.05$) from an expected mean of zero.

8.6   Comparing the means of two independent samples

Often you will need to compare the means of two independent samples. This type of comparison is particularly common when you have two randomly chosen independent samples such as a control and an experimental group, each containing different experimental units. Here the question is "Have the two sample means been drawn from populations with the same mean μ?" which can be tested with an independent sample t test.

It is easy to visualize this pictorially. Under the null hypothesis, each sample is from the same population, so 95% of the time you would expect the two sample means to lie within the 95% confidence interval surrounding μ. Here, however, you are interested in **the range of possible differences between two values of $\overline{X}$**, which will be much wider than the confidence interval for each sample, because there will be cases where one mean is at the lower end of the expected range and the other at the higher end and *vice versa* (Figure 8.6).

To obtain a t statistic for the difference between two independent sample means you simply need to divide $\overline{X}_A - \overline{X}_B$ by the standard error of the distribution of differences shown in Figure 8.6(b). The latter is easy to estimate because the variance of the **difference** between the means of two independent samples is the **sum** of the variances of these samples:

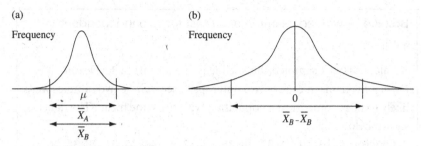

Figure 8.6 Illustration of the comparison made by an independent sample t test. (a) The graph shows the range, indicated by the double-headed arrows, within which 95% of the values of means of samples size n, from a population with a mean of μ, are expected to occur. (b) This graph shows the expected distribution of the **differences** $(\overline{X}_A - \overline{X}_B)$ between any two sample means of size n from that population. The distribution of differences will have a mean of zero (when both $\overline{X}_A$ and $\overline{X}_B$ are equal) and a much greater dispersion than in (a) because there will be cases where $\overline{X}_A$ is at the low end of the range and $\overline{X}_B$ is at the high end of the range (giving large negative values) and vice versa (giving large positive values). The double-headed arrow shows the 95% confidence interval for $\overline{X}_A - \overline{X}_B$. Note that it is much wider than the 95% confidence interval for the sample means shown in (a).

$$S^2{}_{A-B} = S_A{}^2 + S_B{}^2 \tag{8.1}$$

This is consistent with the much greater variance in Figure 8.6(b) compared to Figure 8.6(a). Because the SEM from a sample is $\sqrt{s^2/n}$, you need the best estimate of the standard error of $\overline{X}_A - \overline{X}_B$. So you use the following formula, which is just the square root, of the variance of sample A divided by the sample size of A plus the variance of sample B divided by the sample size of B:

$$\text{SEM} = \sqrt{\frac{s_A{}^2}{n_A} + \frac{s_B{}^2}{n_B}} \tag{8.2}$$

Finally, to obtain the t statistic for the differences between the two means, you divide $\overline{X}_A - \overline{X}_B$ by this estimate of the SEM:

$$t = \frac{\overline{X}_A - \overline{X}_B}{\sqrt{\frac{s_A{}^2}{n_A} + \frac{s_B{}^2}{n_B}}} \tag{8.3}$$

Box 8.4 A worked example of a *t* test for two independent samples

A paleontologist sampled the shell length (in mm) of 15 Devonian-age *Paracyclas* clams in each of two outcrops to see if these two samples were likely to have come from a population with the same mean. The data are shown below:

Outcrop A: 25, 40, 34, 37, 38, 35, 29, 32, 35, 44, 27, 33, 37, 38, 36

Outcrop B: 45, 37, 36, 38, 49, 47, 32, 41, 38, 45, 33, 39, 46, 47, 40

$n_A = 15$, $n_B = 15$, $\overline{X}_A = 34.67$, $\overline{X}_B = 40.87$, $S_A^2 = 24.67$, $S_B^2 = 28.69$

therefore $t_{28} = \frac{34.67 - 40.87}{\sqrt{\frac{24.67}{15} + \frac{28.69}{15}}} = \frac{-6.2}{\sqrt{1.648 + 1.913}} = -3.287$

Note that the value of t is negative because the mean for outcrop B is greater than that of outcrop A.

The critical value of t_{28} for $\alpha = 0.05$ is 2.048, so the two samples have less than 5% probability of being from the same population.

Here the number of degrees of freedom is $(n_{(A)} - 1) + (n_{(B)} - 1)$, which is usually put as $(n_{(A)} + n_{(B)} - 2)$. This is because you have calculated the standard error using two independent samples, both of which have $n - 1$ degrees of freedom, so one degree of freedom is lost from each sample. A worked example is given in Box 8.4.

You may never have to manually calculate a t statistic because statistical packages have excellent programs for doing them. But the simple worked examples in this chapter will help you understand how t tests work and will be very helpful as you continue through this book.

8.7   Are your data appropriate for a *t* test?

The use of a t test makes three assumptions. The first is that the data are normally distributed. The second is that each sample has been taken at random from its respective population and the third is that for an independent sample test, the variances are the same. Of course, in geological applications, you can rarely make these assumptions with impunity. In many cases, distributions and variances are not well-known, and physical constraints such as rock exposure generally make it difficult to take samples from truly random locations.

Fortunately, it has been shown that _t_ tests are actually very "robust" – that is, they will still generate statistics that approximate the _t_ distribution and give realistic probabilities even when the data show considerable departure from normality and when sample variances are dissimilar.

8.7.1   Assessing normality

First, if you already know that the population from which your sample has been taken is normally distributed (perhaps you have data for a variable that has been studied before) you can assume the distribution of sample means from this population will also be normally distributed.

Second, the central limit theorem discussed in Chapter 7 states that the distribution of the means of samples of about 25 or more taken from **any** population will be approximately normal, provided the population is not grossly non-normal (e.g. a population that is bimodal). Therefore, provided your sample size is sufficiently large you can usually do a parametric test.

Finally, you can examine your sample. Although there are statistical tests for normality, many (see Koch and Link, 2002) have cautioned that these tests often indicate the sample is significantly non-normal even when a _t_ test will still give reliable results.

Some authors (e.g. Stanley, 2006) suggest plotting the cumulative frequency distribution of the sample. The easiest way to do this is to use a statistics package to give you a probability plot (often called a P-P plot). This graphs the actual cumulative frequency against the expected cumulative frequency assuming the data are normally distributed. If they are, the P-P plot will be a straight line. Any gross departures from this should be analyzed cautiously and perhaps a non-parametric test used. Most statistical packages will draw a P-P plot for a sample.

8.7.2   Have the sample(s) been taken at random?

This is really just a case of having an appropriate experimental design. For a single-sample test, the sample needs to have been selected at random in order to appropriately represent the population from which it has been taken. For an independent sample test, both samples need to have been selected at random. One potential pitfall associated with this assumption deals with accessibility: it may be tempting to assume that an apparently random outcrop of some geological formation is representative of the unit as a whole. But do not forget that there may be a reason why that particular outcrop is exposed – and that reason may mean that it is unrepresentative. For example, perhaps the outcrop is the remnant of a small basaltic dike more resistant to erosion than the surrounding sandstone, which has now been eroded completely away.

8.7.3   Are the sample variances equal?

One easy test of whether sample variances are equal is to divide the largest by the smallest. If the samples have equal variances, this ratio will be 1.00. As the variances become more and more unequal, the value of this statistic, which is called the **F statistic** or **F ratio** after the statistician Sir Ronald A. Fisher, will increase. There will be discussion of F and tests for equality of variances in Chapters 10 and 12. Even if the variances of two samples are significantly different, you can often still apply a t test.

8.8   Distinguishing between data that should be analyzed by a paired-sample test and a test for two independent samples

As a researcher, or reviewer of another person's work, you may have to decide if an experimental outcome should be analyzed as a paired-sample test or a test for two independent samples. The way to do this is to ask "Are the experimental or sampling units in the two samples related or are they independent?" Here are some examples.

First, Table 8.3 shows two samples that are related. Measurements of wt% FeO have been made on two different size fractions made from each of four units of granite.

Each experimental unit (granite) in Table 8.3 has been ground and sieved to separate the powders into two size fractions ($< 125\ \mu m$ and $125–250\ \mu m$), so you would do a paired-sample test. Here you would be testing whether or not the two size fractions have the same bulk chemistry.

An independent example is the measurement of wt% FeO on four units of granite ground to $45–125\ \mu m$ (granites 1–4) and four more ground to $125–250\ \mu m$ (granites 5–8) shown in Table 8.4. The samples are obviously independent. You would do an independent sample t test.

Table 8.3 Data for the wt% FeO measured on four granites, ground and separated into two different size fractions.

Granite	< 125 μm	125–250 μm
1	0.10	0.13
2	1.50	1.40
3	0.70	0.50
4	1.10	1.20

Table 8.4 Data for the wt% FeO measured on eight granites, ground to different size fractions.

Granite	45–125 μm	125–250 μm
1	1.20	
2	0.50	
3	3.30	
4	1.30	
5		1.35
6		2.60
7		1.20
8		0.40

8.9   Conclusion

This chapter explains how the Z test and t tests for one and two samples actually work. The concepts will help you make decisions about which test to use for a particular set of data and also be very useful when you work through the material in later chapters. They will also help you understand the results given by statistical packages.

8.10   Questions

(1) A geoscientist hypothesized that ^{60}Co irradiation would affect the opacity of amethyst crystals. They measured the opacity of 10 crystals, irradiated them and then remeasured their opacity. The data are shown below, as opacity before and after the irradiation. (a) What sort of

Crystal number	Opacity before	Opacity after
1	3.5	3.7
2	4.6	4.6
3	2.7	2.9
4	5.5	5.7
5	1.1	1.1
6	6.4	6.6
7	3.2	3.1
8	9.3	9.5
9	6.7	6.8
10	8.4	8.5

statistical analysis is appropriate for this hypothesis? Is the hypothesis one or two tailed? Is the result of the analysis significant?

(2) The geoscientist who did the irradiation experiment described in the previous question decided (incorrectly) to analyze their data as a two-sample t test. (a) Analyze the data as two independent samples. What is the result of the analysis? Is it significant? Please comment.

(3) This is a valuable exercise that will help you understand how statistical tests actually work. It can be done by hand using the instructions in Box 8.2, but you can do it very easily indeed if you have access to a statistical package. The water content of a sample of nine vermiculites is 6.2, 5.6, 5.3, 6.8, 6.9, 5.3, 6.3, 6.2 and 5.4 wt% H_2O. The mean of this sample is exactly 6.0. Use a statistical package to run a one-sample t test where the **expected** mean is set at 6.0. This will give no difference between the observed and expected mean. (a) What would you expect the value of the t statistic to be? Run the analysis to check on this. (b) Now, modify the expected value. Make it 5.90 and run the analysis again. Then make it 5.80, 5.75, 5.50, etc. What happens to the value of t as the difference between the observed and expected values increases? What happens to the probability?

9 | Type 1 and Type 2 error, power and sample size

9.1   Introduction

Every time you make a decision based on the probability of a particular result, there is a risk that your decision is wrong. There are two sorts of mistakes you can make and these are called **Type 1 error** and **Type 2 error**.

9.2   Type 1 error

A Type 1 error or **false positive** occurs when you decide the null hypothesis is false when in reality it is not. Imagine you have taken a sample of size n from a population with known statistics of μ and σ and subjected this sample to a particular experimental treatment. Because the population statistics are known you could test whether this sample mean was significantly different to the population mean by doing a Z test (Section 8.3).

If the treatment had no effect the null hypothesis would apply and your sample would simply be equivalent to one drawn at random from the population. Nevertheless, 5% of the sample means of size n will lie outside the 95% confidence interval of $\mu \pm 1.96$ SEM. Therefore, 5% of the time you would incorrectly reject the null hypothesis of no difference between your sample mean and the population mean (Figure 9.1) and accept the alternate hypothesis. This is a Type 1 error.

It is important to realize that Type 1 error can only occur when the null hypothesis applies. There is absolutely no risk if the null hypothesis is false. Unfortunately, you are most unlikely to know if the null hypothesis applies or not – if you did you would not be doing an experiment to test it! If the null hypothesis applies the risk of Type 1 error is the same as the probability level you have chosen.

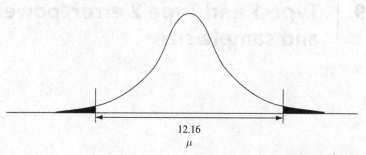

Figure 9.1 Illustration of Type 1 error. The known population mean is 12.16 and the 95% confidence interval for the mean is shown as the double headed horizontal arrow. There is no effect of treatment, so the distribution of sample means from the experimental population will be the same as those from the untreated population. Nevertheless, 5% of your sample means will, by chance, lie in the shaded areas outside the 95% confidence interval. Whenever a sample mean occurs in either of these areas you will incorrectly reject the null hypothesis and make a Type 1 error. This risk is unavoidable when the null hypothesis applies, but can be controlled by the chosen value of α. An α = 0.05 will have a 5% probability of Type 1 error, but an α of 0.01 will only have a 1% probability of Type 1 error.

Here, therefore, you may be thinking "Then why do we usually set α at 0.05? Surely an α of 0.01 or 0.001 would reduce the risk of Type 1 error?" It will, but it will affect the likelihood of Type 2 error.

9.3   Type 2 error

A Type 2 error or **false negative** occurs when you do not reject the null hypothesis even though it is false. For the example above, this would occur when the treatment had a real effect but your experiment and analysis did not detect it. Here is an example, using a single-sample, two-tailed Z test where the population statistics are known.

9.3.1   A worked example showing Type 2 error

The weight percent of Al_2O_3 in lower crustal granulite xenoliths (rocks from deep within the Earth's crust that have become encased in magma and brought to the surface by volcanic activity) has a population mean of 12.16 and a standard deviation of 2.43. These statistics are for a sample size of

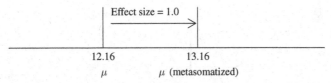

Figure 9.2 The concept of effect size displacing the population mean. The population mean, μ, is 12.16 wt% Al_2O_3 but metasomatism appears to increase the overall Al_2O_3 by 1.0 wt%.

more than a thousand xenoliths, so they can be considered to be the population statistics μ and σ.

You have a sample of seven lower crustal granulite xenoliths that appear to have experienced metasomatism (that is, they have been affected by hydrothermal or other fluids) and wish to determine if this "treatment" has changed their Al_2O_3 content.

Here you need to imagine the case where the metasomatism has caused an average increase in Al_2O_3, so the mean of the metasomatized population is 1.0 wt% more than the mean of the known "untreated" population, but you do not know this. This change is often called the **effect size** of a treatment. To test if this effect is significant, you measure the Al_2O_3 of your seven metasomatized xenoliths and then compare the mean of this sample to that of the untreated population (Figure 9.2).

First, consider the case where you take a sample of $n = 7$ from each population. The expected standard error of each mean will be $\sigma/\sqrt{n} = 2.43/\sqrt{7} = 0.92$. Therefore, the range around μ within which you would expect 95% of sample means from the **untreated population** to occur would be $\mu \pm 1.96 \times$ SEM, which is $12.16 \pm (1.96 \times 0.92)$ and thus 12.16 ± 1.8, giving a wide range from 10.36 to 13.96.

With an effect size of 1.0, the range around μ (metasomatized) within which you would expect 95% of sample means from the **treated population** is 13.16 ± 1.8 which is from 11.36 to 14.96.

These two ranges are shown in Figure 9.3(a). **Importantly, they overlap considerably, with most of the means of samples from the treated population falling within the expected range of the means of samples from the untreated one.** Therefore, if you were to measure seven metasomatized xenoliths, there is a very high probability that their sample mean will fall within the 95% confidence interval of the **untreated** population and thus would not be considered significantly different to μ. **Even though there**

(a)

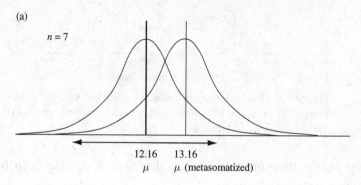

(b)

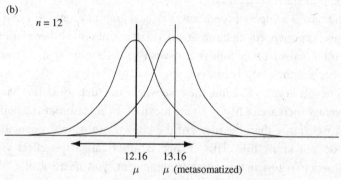

(c)

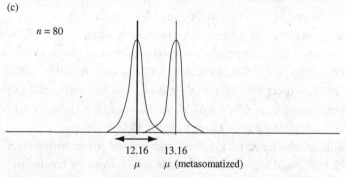

Figure 9.3 Sample size has an effect on the range within which 95% of the means of samples from a population will occur. The expected distributions of the means of samples taken from two populations with the same variance, one of which has a μ of 12.16 and the other which has a μ (metasomatized) of 13.16, are shown. (a) When $n = 7$ the sample means are expected to occur within a relatively wide range around each mean. (b) When $n = 12$ the sample means are expected to occur within a narrower range. (c) When $n = 80$ the sample means are expected to occur within a much narrower range.

is a real effect of metasomatism, your sample size is too small to detect it very often, so you will frequently make a Type 2 error.

Now, consider the case where you have a sample of $n = 12$ metasomatized xenoliths. As sample size increases, the standard error of the mean, and therefore the 95% confidence interval of the mean, will reduce.

For a sample size of 12, the standard error of the mean is $\sigma/\sqrt{n} = 2.43/\sqrt{12} = 0.701$. (Note that this value is smaller than the SEM for the sample of seven given above.) Therefore, the 95% confidence interval for the distribution of values of the mean around μ is 12.16 ± 1.375 (which is from 10.79 to 13.53) and the distribution around μ (metasomatized) is 13.16 ± 1.375 (which is from 11.79 to 14.53). These two ranges are shown in Figure 9.3(b). The confidence intervals have been reduced, but the majority of the sample means from the treated population still lie within the range expected from the untreated one, so the risk of Type 2 error is still very high.

Finally, for a sample size of 80, the standard error will be greatly reduced at $2.43/\sqrt{80} = 0.272$. Therefore, the 95% confidence interval for the mean of a sample of 80 will be $\mu \pm 0.532$, which is from 11.63 to 12.69 for the untreated population, and from 12.63 to 13.69 for the treated one (Figure 9.3(c)). There is little overlap between the 95% confidence intervals of both groups, so you are much less likely to make a Type 2 error. When the sample size is 80, there is only a small risk of failing to reject the null hypothesis that $\mu = 12.16$ wt% Al_2O_3 because only about 5% of the possible values of the sample mean from the treated population are still within the region expected if the mean of 12.16 is correct (Figure 9.4).

The probability of Type 2 error is symbolized by β and is **the probability of failing to reject the null hypothesis when it is false.** Therefore, as shown in Figure 9.4, the value of β is the shaded area of the treated distribution lying to the left of the upper confidence limit for μ.

9.4 The power of a test

The power of a test is the probability of making the correct decision and rejecting the null hypothesis when it is false. Therefore power is the area of the treated distribution to the **right** of the vertical line in Figure 9.4. If you know β, you can calculate power as $1 - \beta$.

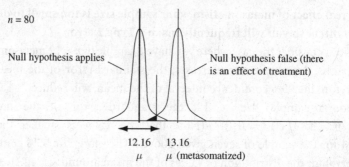

Figure 9.4 The probability of a Type 2 error is the shaded area to the left of the vertical line marking the upper confidence limit (12.69) of μ. The risk of Type 2 error is low, but it will be greater if the sample size is smaller.

An 80% power is considered desirable. That is, there is only a 20% chance of a Type 2 error and an 80% chance of **not** making a Type 2 error when the null hypothesis is false.

9.4.1   What determines the power of a test?

The power of a test depends on several things, only some of which can be controlled by the researcher.

The **uncontrollable factors are effect size and the variance of the population**. As effect size increases, power will increase and will eventually be 100% as the two distributions get further and further apart (Figure 9.5 (a)). Samples from populations with a relatively small variance will have a smaller standard error of the mean, so overlap between the untreated and treated distributions will be less than for samples from populations with a larger variance (Figure 9.5(b)).

The **controllable factors are the sample size and your chosen value of α**. As sample size increases, your risk of Type 2 error decreases and power therefore increases because the standard error of the mean decreases (this has already been described in Figure 9.3).

As the chosen value of α decreases (e.g. from 0.10 to 0.05 to 0.01 to 0.001), the risk of Type 1 error decreases, but the risk of a Type 2 error increases. This is shown in Figure 9.6. There is a trade-off between the risks of Type 1 and Type 2 error.

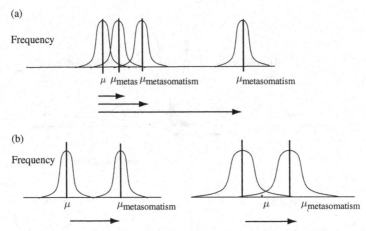

Figure 9.5 Uncontrollable factors affecting power. (a) Effect size will determine power and if the effect size is large enough, power will be 100%. The arrows show effect size. (b) With a fixed effect size, a test comparing the distribution of sample means from a population with a relatively small variance (the pair of graphs on the left) will have greater power than if the population variance is large (the pair of graphs on the right).

9.5   What sample size do you need to ensure the risk of Type 2 error is not too high?

Without compromising the risk of Type 1 error, the only way a researcher can reduce the risk of Type 2 error to an acceptable level and therefore ensure sufficient power, is to increase the sample size. Every researcher has to ask themselves the question "What sample size do I need to ensure the risk of Type 2 error is low and therefore power is high?" This is an important question because samples may be difficult to collect and the characterization may be expensive, so there is no point in increasing sample size past the point where power reaches an acceptable level. For example, if a sample size of 35 gave 100% power, there is no point in taking any more than this number of replicates.

Unfortunately, the only way to estimate the appropriate minimum sample size needed in an experiment is to know, or have good estimates, of the effect size and standard deviation of the population(s) – and this is often impractical, or subject to interpretation, in geological situations. Often the only way to estimate these is to do a pilot experiment with a sample. For most tests there are formulae that use these (sample)

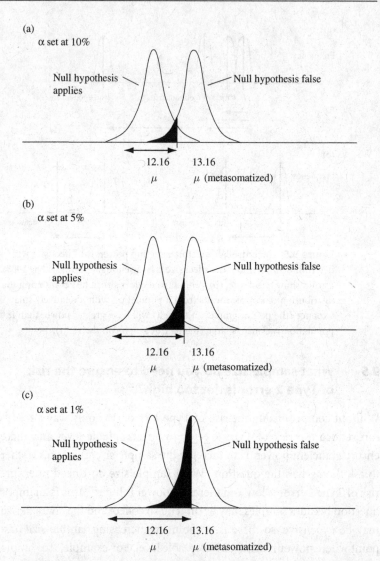

Figure 9.6 The trade-off between Type 1 and Type 2 error. (a) α set at 10%. (b) Decreasing α to 0.05 will reduce the risk of Type 1 error, but will increase the risk of Type 2 error. (c) Decreasing α to 0.01 will further decrease the risk of Type 1 error, but greatly increase the risk of Type 2 error.

statistics to give the appropriate sized sample for a desired power. Some statistical packages will calculate the power of a test as part of the analysis.

9.6   Type 1 error, Type 2 error and the concept of risk

The commonly used α of 0.05 sets the risk of Type 1 error at 5%, while 20% is considered an acceptable risk of Type 2 error. **Nevertheless, these risks have to be considered in relation to the consequences of an incorrect decision about the null or alternate hypotheses.** There was a discussion about the appropriate level risk of Type 1 error depending on the consequences in Chapter 6 and the same considerations apply to the risk of Type 2 error.

For example, a test that has a 20% chance of incorrectly retaining the null hypothesis of no effect may be considered inappropriate if you are testing for the undesirable side effects of a new sample preparation, or evaluating whether the release of lead from a landfill into a river is affecting the concentration of lead in a lake downstream. Every time you run a statistical test you have to consider not only the risk of Type 1 and Type 2 error, but also the consequences of these risks.

9.7   Conclusion

Whenever you make a decision based on the probability of a result, there is a risk of either a Type 1 or a Type 2 error. There is only a risk of Type 1 error when the null hypothesis applies, and the risk is the chosen probability level α. There is only a risk of Type 2 error when the null hypothesis is false. Here the risk of Type 2 error, β, is affected by several factors, but the most controllable is sample size. As sample size increases, the risk of Type 2 error decreases.

Power is the converse of Type 2 error. Power is $1 - \beta$ and is the ability of the test to reject the null hypothesis when it is false.

There are formulae for calculating the appropriate sample size to ensure that the risk of Type 2 error is acceptable (e.g. 20%) and therefore a test has acceptable power, but these calculations rely on an estimate of effect size and the standard deviation of the sample or population.

The risks of Type 1 and Type 2 error also need to be considered in terms of **geoscientific risk.** Depending on the consequences of making each type of error, you may find an α of 5%, or a β of 20%, unacceptable.

Finally, the example of xenoliths used in this chapter deliberately used only one variable and therefore a very specific hypothesis about the effect of metasomatism on Al_3O_2. Often, however, you may be able to measure several variables (i.e. several chemical constituents including Al_3O_2). Therefore, if you are testing the more general hypothesis that "Metasomatism affects the chemical composition of xenoliths" then a multivariate data set will provide more information and may give a more reliable result. Methods for analyzing multivariate data are discussed in Chapter 20.

9.8   Questions

(1) Comment on the following: "Depending on sample size, a non-significant result in a statistical test may not necessarily be correct."
(2) Explain the following: "I did an experiment with only 10% power (therefore β was 90%) but the null hypothesis was rejected so the low power does not matter and I can trust the result."

10 | Single-factor analysis of variance

10.1 Introduction

So far, this book has only covered tests for one and two samples. Often, however, you are likely to have univariate data from three or more samples, from different localities (or experimental groups), and wish to test the hypothesis that "The means of the populations from which these samples have come from are not significantly different to each other," or "$\mu_1 = \mu_2 = \mu_3 = \mu_4 = \mu_5$ etc...."

For example, you might have data for the percentage of tourmaline in granitic rocks from five different outcrops, and wish to test the hypothesis that these have come from populations with the same mean percentage of tourmaline, or perhaps even the same pluton.

Here you could test this hypothesis by doing a lot of two-sample t tests that compare all of the possible pairs of means (e.g. mean 1 compared to mean 2, mean 1 compared to mean 3, mean 2 compared to mean 3 etc.). The problem with this approach is that every time you do a two-sample test and the null hypothesis applies you run a 5% risk of a Type 1 error. So as you do more and more tests on the same set of data, the risk of a Type 1 error rises rapidly.

Put simply, if you do two or more two-sample tests on the same data set it is like having more than one ticket in a lottery where the chances of winning are 5% – the more tickets you have, the more likely you are to win. Here, however, to "win" could be to make the wrong decision about your results. If you have five groups, there are ten possible pairwise comparisons among them and the risk of a getting a Type 1 error when using an α of 0.05 is 40%, which is extremely high (Box 10.1).

Obviously there is a need for a test that compares three or more sample means simultaneously but only has a risk of Type 1 error the same as your chosen value of α. This is where analysis of variance (ANOVA) can often be used.

Box 10.1 The probability of a Type 1 error increases when you make several pairwise comparisons

Every time you do a statistical test where the null hypothesis applies, the risk of a Type 1 error is your chosen value of α. If α is 0.05 then the probability of **not** making a Type 1 error is $(1-\alpha)$ or 0.95.

If you have three means and therefore make three pairwise comparisons (1 versus 2, 2 versus 3 and 1 versus 3) the probability of **no** Type 1 errors is $(0.95)^3 = 0.86$. The probability of **at least** one Type 1 error is 0.14 or 14%.

For four means there are six possible comparisons so the probability of **no** Type 1 errors is $(0.95)^6 = 0.74$. The probability of **at least** one Type 1 error is 0.26 or 26%.

For five means there are ten possible comparisons so the probability of **no** Type 1 error is $(0.95)^{10} = 0.60$. The probability of **at least** one Type 1 error is 0.40 or 40%.

These risks are unacceptably high. You need a test that compares more than two means with a Type 1 error the same as α.

A lot of earth scientists make decisions on the results of ANOVA without knowing how it works. But it is very important to understand how ANOVA does work so that you can appreciate its uses and limitations!

Analysis of variance was developed by the statistician Sir Ronald A. Fisher from 1918 onwards. It is a very elegant technique and can be applied to numerous and very complex experimental designs. This book introduces the simpler ANOVA models because an understanding of these makes the more complex ones easier. The following is a pictorial explanation, like the ones developed to explain t tests in Chapter 8. This approach is remarkably simple and does represent what happens. By contrast, a look at the equations in many statistics texts makes ANOVA seem very confusing indeed.

10.2   Single-factor analysis of variance

Imagine you are interested in understanding the occurrence of tourmaline in the pegmatites scattered throughout western Maine. This area was the source of the first gem tourmaline mined in the US, which was discovered at Mount Mica (just outside of Paris, Maine) in 1820. Subsequent exploration has found several other pegmatites, some of which have been mined for industrial minerals, including gemstone varieties of the tourmaline group.

However, not all pegmatites are the same, apparently because the parent magmas have different chemistries. Some contain valuable green, pink and two-tone ("watermelon") gemmy tourmalines, but others have only the glossy black elongated crystals of the schorl species.

Prospecting to discover new gem-containing pegmatites in the region would be greatly simplified if the genetic relationships among the existing ones could be clarified. One way of distinguishing among pegmatites is to measure the ratio between the stable isotopes of oxygen, ^{18}O and ^{16}O in tourmalines. The results are reported in "delta" notation as $\delta^{18}O$ per mil (‰) units relative to $\delta^{18}O$ in Vienna Standard Mean Ocean Water (VSMOW: previously discussed in Chapter 8).

You have obtained isotopic data on samples of tourmaline from three different localities. In statistical terms, these three localities represent, and are often called, different **treatments**. At each location four tourmalines were collected. In statistical terms these are called **replicates** and correspond to the sampling units described in Chapter 1. The total number of replicates from each location comprises a **sample**.

A sample of four tourmalines was collected from the Sebago Batholith, the largest pluton in Maine and the possible "parent" magma body for smaller occurrences.

Another sample of four was collected from the Mount Mica pegmatite, which is a shallowly dipping sill of undetermined thickness located ~4 km to the northeast of the Sebago Batholith.

The final sample of four specimens was from the Black Mountain pegmatite in Rumford, ~15 km north of the Sebago Batholith.

Your null hypothesis is that "There is no difference in isotopic composition among the populations from which these three samples have been taken." The alternative hypothesis is "There is a difference in isotopic composition among the populations from which these samples have been taken."

The results of this sampling have been displayed pictorially in Figure 10.1, with $\delta^{18}O$ increasing on the Y axis and the three treatment categories on the X axis. The sample means of each group of four are shown, together with the grand mean, which is the mean $\delta^{18}O$ of all 12 tourmalines.

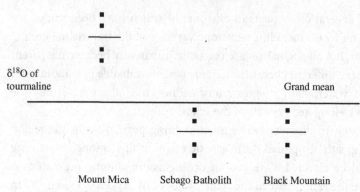

Mount Mica   Sebago Batholith   Black Mountain

Figure 10.1 Pictorial representation of the oxygen stable isotope ratio for tourmalines from three localities in Maine. The value of $\delta^{18}O$ for tourmaline increases up the page. The heavy horizontal line shows the grand mean, while the shorter lighter lines show the means for each location. The value for each replicate tourmaline analysis is shown as a filled square ■.

Now, think about the data for each tourmaline. There are two possible sources of variation that will contribute to its displacement from the grand mean.

First, there is the effect of the locality (i.e. the treatment) it is from (the Sebago Batholith, Mount Mica or Black Mountain).

Second, there is likely to be variation **within** each of these three deposits that cannot be controlled, such as slight differences in cooling history, heterogeneity of the magma, and interactions with groundwater, plus errors associated with the isotopic measurements. This uncontrollable variation is called "error."

Therefore, the displacement of each point on the Y axis from the grand mean will be determined by the following formula:

$$\delta^{18}O \text{ of tourmaline} = \text{treatment} + \text{error} \qquad (10.1)$$

In Figure 10.1, tourmalines from the Sebago Batholith and Black Mountain appear to be similar (so perhaps they are co-genetic), while Mount Mica seems to have a distinctly higher $\delta^{18}O$ value, but **is this significant, or is it just the sort of difference that might occur by chance among samples taken from populations with the same mean?** A single factor ANOVA calculates this probability in a very straightforward way. The key to understanding how the ANOVA does this is to consider the reasons why the values for each replicate and the treatment means are where they are.

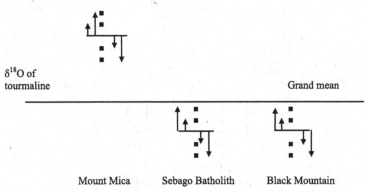

$\delta^{18}O$ of
tourmaline

Grand mean

Mount Mica   Sebago Batholith   Black Mountain

Figure 10.2 Arrows show the displacement of each replicate from its respective treatment mean. This is the variation due to error only.

First, the isotope results for the four individual tourmalines from each location will be displaced from the treatment mean by error only. This is called **error** or **within group variation** (Figure 10.2).

Second, each treatment mean will be displaced from the grand mean by **any effect of that treatment plus error**. Here, because we are dealing with treatment means, the distance between a particular treatment mean and the grand mean is the average effect of all of the replicates within that treatment. To get the total effect you have to think of this displacement occurring for each of the replicates. This is called **among group variation** (Figure 10.3).

Third, the stable isotope ratio for each of the 12 tourmalines will be displaced from the grand mean by **both sources of variation** – the within group variation (Figure 10.2) plus the among group variation (Figure 10.3) described above. This is called the **total variation**. In Figure 10.4 the distance displaced is shown for the four tourmalines in each treatment.

Figures 10.2 to 10.4 show the dispersion of points around means. Therefore it is possible to calculate separate variances from each figure.

(a) **The within group variance, which is due to error only** (Figure 10.2) can be calculated from the dispersion of the replicates around each of their respective treatment means.

(b) **The among group variance, which is due to treatment and error** (Figure 10.3) can be calculated from the dispersion of the treatment means around the grand mean. The distance between each treatment

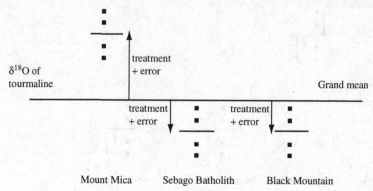

Figure 10.3 The arrows show the displacement of each treatment mean from the grand mean and represent the average effect of the treatment plus error for the replicates in that treatment.

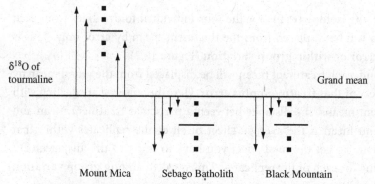

Figure 10.4 Arrows show the displacement of each replicate from the grand mean. The length of each arrow represents the total variation affecting each replicate.

mean and the grand mean will represent the average effect for the number of replicates in that treatment.

(c) **The total variance** (Figure 10.4) is the combined effects of the within group variance and the among group variance (quantities "a" and "b" above). This can be calculated from the dispersion of all the points around the grand mean.

These estimates give you a very useful way of assessing whether the three treatment means have come from populations with the same mean μ.

First, **if there is no effect of any treatment** (in this case each pegmatite), the among group variance (due to treatment plus error) will be a small

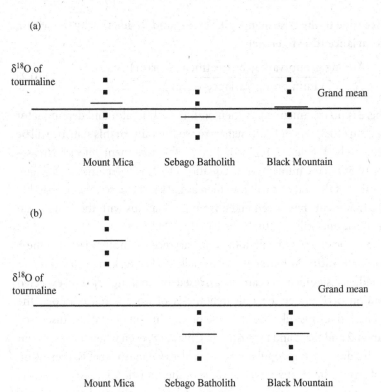

Figure 10.5 Pictorial representation of (a) No effect of treatment. The three treatment means are only displaced from the grand mean because of error, so the "among group" variance will be relatively small. (b) An effect of treatment. There are relatively large differences among the treatment means, so they are further from the grand mean causing the among group variance to be relatively large.

number because all the treatment means will only be displaced from the grand mean by any effect of error (Figure 10.5(a)).

Second, **if there is a relatively large treatment effect**, some or all of the treatment means will be very different to each other and further away from the grand mean. Therefore the among group variance (due to treatment plus error) will be large compared to the within group variance (due to error only) (Figure 10.5(b)). As the differences among treatments get larger and larger so will the among group variance.

Therefore, to get a statistic that shows the **relative effect of the treatments compared to error, all you have to do is calculate the among group** variance (due to the treatments plus error) and divide this by the within group variance (due to error):

$$\frac{\text{Among group variance (treatment + error)}}{\text{Within group variance (error)}} \qquad (10.2)$$

If there is no treatment effect then both the numerator and denominator of Equation (10.2) will only estimate error so the value of this statistic will be approximately 1 (Figure 10.5(a)). But as the treatment effect increases (Figure 10.5(b)), the numerator of Equation (10.2) will get larger and larger, so the value of the statistic will also increase. As it increases, the probability that the treatments have been taken from populations with the same mean will **decrease** and will eventually be less than 0.05.

The statistic obtained by dividing one variance by another is called the **F statistic** or **F ratio**, in honor of Sir Ronald A. Fisher. Once an F ratio is calculated, its significance can be assessed by looking up the expected distribution of F under the null hypothesis of no difference among the treatment means. Just like the example of the chi-square statistic discussed in Chapter 2 and the Z and t statistics in Chapter 8, even when the treatment groups are drawn from populations with the same mean (that is, there is no effect of any of the treatments) the value of the statistic will, just by chance, be larger than a particular value in 5% of cases and can be considered statistically significant.

10.3   An arithmetic/pictorial example

Doing a single-factor analysis of variance is straightforward and the following example will also help you interpret the results provided by statistics programs. Here we will return to the example of the Maine pegmatites, but will use a different variable to assess the possible differences among localities: the amount of magnesium in the tourmaline, expressed in terms of weight % MgO. We are using a simplified set of data for tourmalines sampled at three localities (treatments), with each of these three samples containing four replicates (Table 10.1).

To do a single-factor ANOVA, all you have to do is calculate the among group (treatment) variance and divide this by the within group (error) variance to get the F ratio. The procedure is shown pictorially below.

Table 10.1 The weight percent of MgO present in tourmalines from (a) Mount Mica, (b) the Sebago Batholith, and (c) Black Mountain.

Mount Mica	Sebago Batholith	Black Mountain
7	4	1
8	5	2
10	7	4
11	8	5

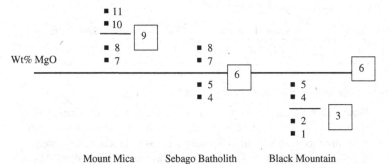

Mount Mica   Sebago Batholith   Black Mountain

Figure 10.6 Pictorial representation of the MgO content of tourmalines from three localities in western Maine, expressed in terms of weight percent MgO content which increases with distance up the page. The heavy horizontal line shows the grand mean, while the shorter lighter lines show treatment means. The wt% MgO content of each replicate is shown as ■. Boxes show the values of the three treatment means and the grand mean.

10.3.1 Preliminary steps

First, you calculate the grand mean, by taking the sum of all the values, and dividing this by n (which is 12). The value of the grand mean is shown in the large box to the right of the line indicating the position of the grand mean in Figure 10.6.

Second, you calculate each treatment mean, by taking the sum of the values in each treatment and dividing by the appropriate sample size (here, in each case it is 4). These values are shown in the boxes to the right of the lines indicating each treatment mean.

These are all the values you need to calculate the three different variances.

Figures 10.7, 10.8 and 10.9 show the calculation of the total, error and treatment variances. The general formula for any sample variance is:

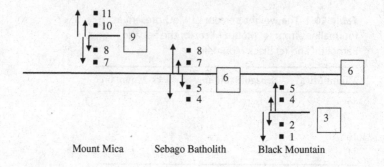

Step 1: The within group (error) **sum of squares** is:

Mount Mica					Sebago Batholith					Black Mountain					Sum of squares
4	1	1	4	+	4	1	1	4	+	4	1	1	4	=	30

Step 2: The within group (error) **variance** is $30 \div 9 = 3.33$

Figure 10.7 Calculation of the within group (error) sum of squares and variance. This has been done in two stages. First, the displacement of each point from its treatment mean has been squared and these values added together to get the sum of squares. Second, this value has been divided by the number of degrees of freedom to give the mean square value, which is the within group (error) variance.

$$\sum \frac{(X_i - \bar{X})^2}{n - 1} \qquad (10.3)$$

and the variances have been calculated in two steps. First the sum of each value minus the appropriate mean and then squared (the numerator of the equation above which is called the **sum of squares**) has been calculated. Second this value has been divided by the appropriate degrees of freedom (the denominator of the equation above) to give the variance, which is often called the **mean square**.

10.3.2 Calculation of within group variation (error)

This has been done in two steps in Figure 10.7. First, you calculate the sum of squares for error. The distance between each replicate and its treatment mean is the error associated with that replicate. You square each of these values and add them together to get the sum of squares.

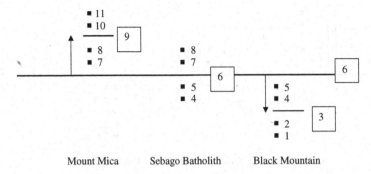

Mount Mica　　　　Sebago Batholith　　　　Black Mountain

Step 1: The total among group (treatment) **sum of squares** is the sum of the average displacement of each treatment, squared and multiplied by the sample size of each treatment:

Mount Mica　　Sebago　　　　　Black Mtn.　　Sum of squares

$\boxed{9} \times 4 + \boxed{0} \times 4 + \boxed{9} \times 4 = \boxed{72}$

Step 2: The among group (treatment) **variance** is $72 \div 2 = 36$

Figure 10.8 Calculation of the among group (treatment) sum of squares and variance. This has been done in two steps. First, the displacement of each treatment mean from the grand mean has been squared. This value has to be multiplied by the sample size within each treatment to get the total effect for the replicates within that treatment because the displacement is the average for the treatment. These three values are then added together to give the sum of squares. Second, this value has been divided by the number of degrees of freedom to give the mean square value, which is the among group (treatment) variance. Note that one of the treatment means happens to be the same as the grand mean, but this will not always occur.

Second, you calculate the error variance (often called the error mean square) by dividing the total by the degrees of freedom. To obtain the appropriate number of degrees of freedom you need to take one away from the number within each treatment and sum the degrees of freedom remaining. Because each treatment contains four replicates the number of degrees of freedom is $3 + 3 + 3 = 9$.

10.3.3　Calculation of among group variation (treatment)

This has been done in two steps in Figure 10.8. First, you calculate the sum of squares for treatment. The distance between any of the three treatment means

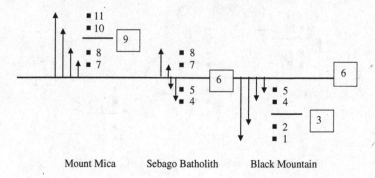

Step 1: The total **sum of squares** is:

Mount Mica					Sebago Batholith					Black Mountain				Sum of squares
25	16	4	1	+	4	1	1	4	+	25	16	4	1	=	102

Step 2: There are 11 degrees of freedom, so the total **variance** is: $102 \div 11 = 9.273$

> **Figure 10.9** Calculation of the total sum of squares and total variation. This has been done in two steps. First, the displacement of each point from the grand mean has been squared and these values added together to give the sum of squares. Second, this value has been divided by the number of degrees of freedom to give the mean square, which is the total variance.

and the grand mean is the average effect of that treatment. Therefore, to get the total effect for all the replicates within each treatment, this value has to be squared **and then multiplied by the number of replicates in that treatment** and these values added together to give the sum of squares for treatment.

Second, you calculate the variance (often called the mean square) by dividing the sum of squares by the degrees of freedom, which is $n - 1$ where n is the number of treatments. Here, because there are three treatments, there are only two degrees of freedom.

10.3.4 Calculation of the total variation

First you calculate the sum of squares for the total variation by taking the displacement of each point from the grand mean, squaring it and adding these together for all replicates. This gives the total sum of squares. Dividing by the number of degrees of freedom (because this is a sample of 12, there are $n - 1$ degrees of freedom which in this case is 11) gives the mean square. This has been done in two steps in Figure 10.9.

Table 10.2 Summary of the results of the calculations from Figures 10.7 to 10.9. The results have been formatted as a typical single-factor ANOVA summary table provided by most statistical software packages. Note the significant probability of 0.004.

Source of variation	Sum of squares	df	Mean square	F ratio	Probability
Among groups (treatment)	72	2	36.0	10.8	0.004
Within groups (error)	30	9	3.3		
Total	102	11			

Finally, to obtain the F ratio, which compares the effect of treatment to the effect of error, you simply divide the **among group (treatment) variance** by the **within group (error) variance**.

Because the treatment variance is 36 (Figure 10.8) and the error variance is 3.33 (Figure 10.7), the F ratio of treatment variance/error variance is $36/3.33 = 10.8$. Table 10.2 gives the results of this analysis in a similar format to the one provided by many statistical software packages.

Here you may be wondering why the total sum of squares and total variance in the experiment have been calculated because they are not needed for the F ratio given above. The calculation has been included to illustrate the additivity of the sums of squares and degrees of freedom. Note from Table 10.2 that the total sum of squares (102) is the sum of the treatment (72) plus the error (30) sums of squares. Note also that the total degrees of freedom (11) is the sum of the treatment (2) plus the error (9) degrees of freedom. This additivity of sums of squares and degrees of freedom will be used when discussing more complex ANOVA models.

Now, all you need is the critical value of the F ratio. This used to be a tedious procedure because there are two values of the degrees of freedom to consider – the one associated with the treatment mean square and the one associated with the error mean square – and you had to look up the critical value in a large set of tables. Here, however, you can use a statistics program to run this analysis, generate the F ratio and obtain the probability. There is a significant difference among the three treatments because the probability (0.004) given in the column on the far right of Table 10.2 is less than 0.05.

The F ratio is always written with the number of degrees of freedom for the numerator and denominator given in order as a subscript. Therefore the F ratio for the among group mean square divided by the within group mean square from Table 10.2 would be written as $F_{2,9}$ because there are two degrees of freedom for the among group variance and nine degrees of freedom for the within group variance.

10.4   Unequal sample sizes (unbalanced designs)

The examples in this chapter have used a sampling design with equal numbers in each treatment. If they are not equal, the method for calculating the F ratio will still work, but the means and variance within each group will not be estimated with the same precision (Chapter 7). For example, the mean of a relatively small sample is likely to be less precise than that of a larger one, so the conclusion from a comparison of means may be misleading. You should, wherever possible, aim to have equal numbers in each treatment especially when sample sizes are relatively small.

10.5   An ANOVA does not tell you which particular treatments appear to be from different populations

Although a significant result of a single-factor ANOVA indicates that the treatments are unlikely to come from populations with the same mean, it has not shown where the differences actually lie. In the example given above, a significant effect might be caused by all three pegmatites having different percentages of MgO; or by one having a significantly higher percentage than the other two which were lower and similar, or by one having a significantly lower percentage than the other two which were higher and similar. You will almost certainly want to know at least which pegmatite has the highest, and which one the lowest percentage of MgO. To do this, you will need to make multiple comparisons among the treatment means. This procedure is described in Chapter 11.

10.6   Fixed or random effects

This is an important concept. There are two types of single-factor ANOVA, which are called Model I and Model II. An understanding of the difference between them is necessary, particularly when you meet two-factor ANOVAs later in this book.

A Model I or **fixed effects** ANOVA applies when the treatments (e.g. the three localities) have been **specifically chosen**. You are only interested in comparing three pegmatites and the null hypothesis reflects this – "There is no difference in MgO content of pegmatites from Mount Mica, Sebago Batholith and Black Mountain."

A Model II or **random effects** ANOVA applies to more general hypotheses. Instead of only comparing these specific localities the hypothesis might be "There is no difference in MgO content among pegmatites in Maine." Therefore the three localities chosen and used in the experiment are merely **random representatives** of all the pegmatites that occur in Maine.

For a single-factor ANOVA the actual computations for both models are the same. But if you have done a Model II ANOVA you would not normally go any further and make multiple comparisons among treatments because you would not be interested in knowing which of the randomly chosen treatments were different. This is discussed in more detail in Chapter 11. When you do two-factor ANOVAs, which are discussed in Chapter 12, it also matters whether the effects are fixed or random.

10.7 Questions

(1) The following simple set of data is for three "treatment" groups, each of which contains four replicates: Treatment A: 1, 2, 3, 4; Treatment B: 1, 2, 3, 4; Treatment C: 1, 2, 3, 4. The mean of each group is the same. The data give some within group (error) variance around each treatment mean, but because the treatment means are identical there is no variation among groups. (a) Do you expect the **within group** (error) sum of squares and mean square values to be zero? (b) Do you expect the **among group** sum of squares and mean square values to be zero? Use a statistical package to run a single-factor ANOVA on these data. (c) Are the results consistent with what you expected? Finally change the values for one treatment group to 21, 22, 23 and 24, run the analysis again and look at the mean square values and F ratio. (d) Is there a significant difference among groups? (e) Have the within group (error) sum of squares and mean square changed from the analysis in (c)? Can you explain this?

(2) Which of the following experimental designs may be suitable for analysis as a Model I ANOVA? (a) A geoscientist was interested in testing the general hypothesis that the mean grain size of sediment varies among alpine lakes. They selected three lakes (Lake Veronica, Lake Michael and Lake Monica) at random from a total of 21 lakes, and took a sample of 10 replicates from within each. (b) A geoscientist was interested in testing the specific hypothesis that the mean grain size of sediment varied among three alpine lakes (Lake Veronica, Lake Michael and Lake Monica), so they took a sample of 10 replicates from within each. (c) A petroleum geologist was asked to analyze whether the mean daily yield of oil differed significantly among the only six offshore wells owned and operated by the Sando Oil Company in the Gulf of Mexico and identify whether any well(s) gave significantly higher yields.

(3) An eminent geographer recently said "The concept of Model I and Model II is irrelevant to single-factor ANOVA, because the calculations are the same in each case." Do you agree or disagree? Why?

(4) An earth scientist did a single-factor ANOVA and obtained a treatment F ratio of 0.99. They said "That F ratio isn't significant. There isn't even a need to look up a table of probability values." One of their colleagues was very worried by that and said "I think you had better look up the probability! You can't be so sure!" Who was right? Why?

11 | Multiple comparisons after ANOVA

11.1   Introduction

When you use a single-factor ANOVA to examine the results of a mensur-ative or manipulative experiment with three or more samples or treatments, a significant result only indicates that one or more appear to come from populations with different means. **It does not identify which particular treatment means appear to be from the same or different populations.**

A significant difference among the means of the three treatments A, B and C can occur in several ways. Mean A may be greater (or less) than B and C; mean B may be greater (or less) than A and C; mean C may be greater (or less) than A and B, and finally means A, B and C may all be different to each other. For example, in Chapter 10 we discussed data for the $\delta^{18}O$ of pegmatites from three locations in Maine. A single-factor ANOVA will only tell you whether (or not) there is a significant difference in $\delta^{18}O$ among these three locations.

If the treatments have been chosen as random representatives of all the possible treatments available (i.e. the factor is random so you have done a Model II ANOVA), then you will not be interested in knowing which particular treatment means appear to be from the same or different popula-tions because your hypothesis is more general. A significant result will reject the null hypothesis and show a difference, but that is all you will want to know.

In contrast, if the treatments have been specifically chosen (i.e. the factor is fixed so you have done a Model I ANOVA) you will be interested in knowing which treatment means appear to be from the same or different populations. There are several multiple comparison tests designed to do this.

11.2   Multiple comparison tests after a Model I ANOVA

Multiple comparison tests are used to make comparisons among a set of means and assign them to groups that appear to be from the same population.

These tests are usually done after a Model I ANOVA has shown a significant difference among treatments. They are called **a posteriori** or **post hoc** tests, both of which mean "after the event," where the "event" is a significant result of the ANOVA.

A lot of multiple comparison tests have been developed but all of them work in essentially the same way. Here is an example using the Tukey test, which works in an analogous way to the two-sample t test described in Chapter 8. The t statistic is calculated by dividing the difference between two means by the standard error of that difference. In contrast, the Tukey statistic, q, is calculated by dividing the difference between two means by the standard error of the mean. The smaller mean is always taken away from the larger, therefore giving a positive number:

$$q = \frac{\bar{X}_A - \bar{X}_B}{\text{SEM}} \tag{11.1}$$

This procedure is first used to compare the largest mean to the smallest. If the difference is significant, testing continues by comparing the largest with the next smallest and so on. If a non-significant difference is found, all the means included within the range between that pair are assigned to the same population. Then the procedure is repeated, starting with the second largest and the smallest mean; repeated again starting with the third largest and the smallest mean, and so on. Eventually the means will be assigned to one or more groups, each containing those which appear to be from the same population (Figure 11.1).

From the example in Figure 11.1, means A, B and C appear to be from the same population and D and E from a second population. The analysis has revealed two distinct groups.

For the Tukey statistic, you need the SEM and the best way to obtain this is from the error mean square of the ANOVA which is an estimate of the population variance, σ^2, calculated from the displacement of **all** the replicates in the experiment from their respective treatment means. Therefore, because the standard error of a mean is:

$$\text{SEM} = \frac{\sigma}{\sqrt{n}} \quad \text{or} \quad \sqrt{\frac{\sigma^2}{n}} \tag{11.2}$$

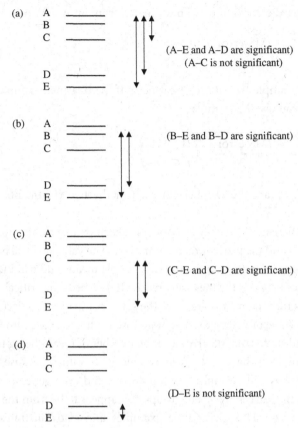

Figure 11.1 General procedure for a Tukey a posteriori test. The treatment means (A to E) are displayed in order of magnitude from the smallest (E) to the largest (A). (a) First the largest mean is compared to the smallest (A–E). If the difference is significant, the largest is then compared to the second smallest (A–D) and so on, until a non-significant difference (here, as an example, A–C) is found or there are no more pairs of means left to compare. All means included within the range between A–C (A, B and C) are assigned to the same population. (b) Testing continues using the same procedure but starting with the second largest mean and comparing it to the smallest (B–E). (c) The third largest mean (C) is compared to D and E. (d) The fourth largest (D) is compared to E. This difference is not significant so D and E appear to be from the same population, which has a different mean to the one from which A, B and C have been taken.

then the standard error of the mean estimated from an ANOVA is:

$$\text{SEM} = \sqrt{\frac{\text{MS error}}{n}} \tag{11.3}$$

where n is the sample size of each treatment. If the treatment sample sizes are different, you use the formula:

$$\text{SEM} = \sqrt{\frac{\text{MS error}}{2} \times \left(\frac{1}{n_A} + \frac{1}{n_B}\right)} \tag{11.4}$$

where n_A and n_B are the numbers in each of the two treatments being compared.

Then you calculate the Tukey statistic q for each pair of means by using Equation (11.1) and the procedure in Figure 11.1. The value of q will be zero when there is no difference among the two sample means and will increase as the difference between the means increases. If q exceeds the critical value, the hypothesis that the means are from the same population is rejected.

The critical value of q depends on your chosen value of α, the number of degrees of freedom for the MS error and the number of means being tested. Here we deliberately have not given a table of q values because most statistical packages will do multiple comparisons and even generate a display assigning the sample means to groups that appear to be from the same population. Section 11.3 gives three examples and also illustrates that ambiguous results are possible.

11.3   An a posteriori Tukey comparison following a significant result for a single-factor Model I ANOVA

11.3.1   Trace elements in New England granites

Trace elements (especially the rare earth elements, Zr, Hf, Ta, Sc and Th) are very useful for understanding crystallization histories and origins of granitic magmas. The relative abundances of these elements are often used as "fingerprints" to determine if geographically separated granite outcrops have come from the same parent magma. Table 11.1 gives data for the hafnium (Hf) contents of four different granitic bodies in New England. This is a Model I ANOVA, because the researcher is only interested in the granite at these four locations.

Table 11.1 Hf contents in (µg/g) of four different New England granites.

Cape Dan	Seabody	Wincy	Easterly
16	19	25	12
14	20	30	10
16	22	26	12
17	20	27	13
18	24	28	9
$\bar{X}$ 16.2	21.0	27.2	11.2

Table 11.2 $\delta^{18}O$ values for tourmalines from three localities in western Maine: the Sebago Batholith, Black Mountain and Mount Mica.

	Sebago Batholith	Black Mountain	Mount Mica
	12.2	12.4	14.8
	13.1	12.6	13.2
	12.7	12.1	13.5
	12.6	11.9	14.6
$\bar{X}$	12.65	12.25	14.03

If you run a single-factor ANOVA on the data in Table 11.1 you will obtain an F ratio ($F_{3,16}$) of 74.01, which has a probability of less than 0.001. Therefore, at least some of the treatment means appear to be from different populations. If you then run an a posteriori Tukey test you will find that all four means appears to be from four distinctly different populations.

11.3.2 Stable isotope data from tourmalines in Maine

Table 11.2 gives data for the $\delta^{18}O$ values for tourmalines from three localities in western Maine: the Sebago Batholith, Black Mountain and Mount Mica. Here too the mean $\delta^{18}O$ values can be used to help decide whether the tourmalines have originated from the same or different parent magmas. A single-factor ANOVA will give an F ratio ($F_{2,9}$) of 12.06, which has a probability of 0.003. At least two treatment means do not appear to be from the same population.

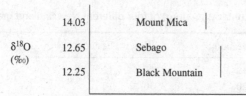

Figure 11.2 Summary of the results of an a posteriori Tukey test comparing among the means of the three samples in Table 11.2. Treatment means connected by vertical lines are not significantly different.

If you run an a posteriori Tukey test it will show that means for the Sebago Batholith and Black Mountain appear to be from the same population, while the mean for Mount Mica appears to be from another with a significantly greater $\delta^{18}O$ value (Figure 11.2).

11.3.3 Apatite in sandstone

The percentage of apatite in sandstone shows considerable variation and can be used to determine the source areas for these sediments. Table 11.3 gives data for the modal abundance of apatite at three different locations. A single-factor ANOVA analysis gives an F ratio ($F_{2,9}$) of 10.8, which has a probability of 0.004. The three treatment means do not appear to be from the same population.

If, however, you run an a posteriori Tukey test it will show that means for Darcy and Runcan appear to be from the same population, while the means for Runcan and Alinda appear to be from another.

This result (Figure 11.3) is obviously ambiguous. The a posteriori analysis has separated the data into two subsets, but the mean of the Runcan sandstone cannot be distinguished from the means of either the Darcy or the Alinda sandstone. At the same time, the mean of the Darcy can be distinguished from the mean for Alinda. Therefore, it seems at least one Type 2 error has been committed somewhere because the mean of the Runcan sandstone has been assigned to two different populations. This is a common problem and is discussed in more detail in the following two sections.

11.3.4 Power and a posteriori testing

Chapter 10 began with a discussion about the danger of an increased probability of Type 1 error when making numerous pairwise comparisons

Table 11.3 The modal percentage of apatite in sandstones from three different basins.

	Darcy	Runcan	Alinda
	7	4	1
	8	5	2
	10	7	4
	11	8	5
$\bar{X}$	9.0	6.0	3.0

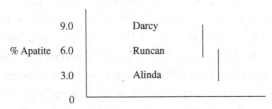

9.0	Darcy	
% Apatite 6.0	Runcan	
3.0	Alinda	
0		

Figure 11.3 Summary of the results of an a posteriori Tukey test comparing among the means of the three samples in Table 11.3. Treatment means connected by vertical lines are not significantly different. The test has assigned the mean for Runcan into two groups (Darcy/Runcan) and (Runcan/Alinda) so at least one Type 2 error has occurred.

among three or more means. Here, however, the a posteriori method for identifying which treatment means appear to be from the same population uses numerous pairwise comparisons. Therefore, you may be thinking that this procedure will also have an increased risk of Type 1 error.

First, however, unplanned a posteriori comparisons are usually only made across all groups if the ANOVA has detected a significant difference among the treatment means. Second, a posteriori tests are specifically designed to take into account the number of means being compared and have a much lower risk of Type 1 error than the same number of t tests. Unfortunately this makes multiple comparison tests relatively low in power, which is why they can give ambiguous results such as the one in Section 11.3.3. In more extreme cases it sometimes happens that an ANOVA detects a significant difference, but subsequent a posteriori testing fails to detect a significant difference among any means. One solution is to increase the sample size of each group or treatment (Chapter 8).

11.4   Other a posteriori multiple comparison tests

There are many other multiple comparison tests. These include the LSD, Bonferroni, Scheffe and Student–Newman–Keuls. The most commonly used are the Tukey and Student–Newman–Keuls (Zar, 1996). Most statistical packages offer you a wide choice of these tests, the relative merits of which are discussed in more advanced texts.

11.5   Planned comparisons

Instead of making a large number of indiscriminate unplanned a posteriori comparisons, a better approach can be to make a small number of more careful (a priori meaning "before the event") comparisons. For example, in Section 11.3.2 you may have a good reason based on outcrop appearance or geographical proximity to propose the following two a priori hypotheses: "Oxygen isotopic ratios of tourmalines at Black Mountain are significantly different than those at Mount Mica" and "Oxygen isotopic ratios of tourmalines at Black Mountain are significantly different than those at the Sebago Batholith." An ANOVA will test for differences among treatments with an α of 0.05 and also give a good estimate of the sample variance from the MS error, since this has been calculated from all the individuals used for this overall comparison. Next, however, instead of making a large number of unplanned comparisons, you could carry out two t tests comparing the mean oxygen isotopic ratio at Black Mountain and Mount Mica, and Black Mountain and the Sebago Batholith.

If you make only one planned comparison the probability of Type 1 error is an acceptable 0.05. If you make several a priori comparisons that **really have been planned for particular reasons before the experiment** (e.g. the two listed above), then each is a distinct and different hypothesis, so the risk of a Type 1 error is still an acceptable 0.05. It is only when you make indiscriminate comparisons that the risk of Type 1 error increases and you should consider using one of the a posteriori tests described previously, which maintains an α of 0.05.

To make a planned comparison after a single factor ANOVA you use the formula for a t test from Chapter 8 except that you use the mean square error as the best estimate of s^2:

$$t_{nA+nB-2} = \frac{\bar{X}_A - \bar{X}_B}{\sqrt{\frac{MS\ error}{1} \times \left(\frac{1}{n_A} + \frac{1}{n_B}\right)}} \qquad (11.5)$$

which reduces to Equation (11.6) when there are equal numbers in both treatment groups:

$$t_{nA+nB-2} = \frac{\bar{X}_A - \bar{X}_B}{\sqrt{\frac{2 \times MS\ error}{n}}} \qquad (11.6)$$

Here is an example, using the data from Example 2 in Section 11.3.2.

From the ANOVA, the mean square for error is 0.288. The mean of the Black Mountain tourmaline $\delta^{18}O$ is 12.25 ‰ and Mount Mica is 14.03 ‰. Therefore:

$$t_6 = \frac{12.25 - 14.03}{\sqrt{\frac{2 \times 0.288}{4}}} = -4.69$$

From Table 8.1 the critical two-tailed 5% value for t is ± 2.447. The two means appear to be from different populations. This supports the idea that these two pegmatites crystallized from unrelated magmas.

The planned comparison to test the Sebago Batholith tourmaline compared to Black Mountain is:

$$t_6 = \frac{12.65 - 12.25}{\sqrt{\frac{2 \times 0.288}{4}}} = 1.05$$

The critical two-tailed 5% value for t is ± 2.447, so these two means do not appear to be from different populations.

Although you are likely to examine other different types of chemical data to further test this conclusion, it appears plausible that the Black Mountain and Mount Mica pegmatites do not share parent magmas, but the Sebago Batholith and the Black Mountain pegmatite do originate from the same parent. Note that this result is consistent with the Tukey test of these data in Section 11.3.2.

Finally, here are two planned comparisons applied to the data for apatite content of sandstones in Table 11.3. Your a priori hypotheses are: "The percentage of apatite in Runcan sandstone is different from that in Darcy" and "The percentage of apatite in Runcan sandstone is different from that in Alinda." Again, the MS error from an initial ANOVA will give a good estimate of the variance.

For the comparison between Runcan and Darcy:

$$t_6 = \frac{6.00 - 9.00}{\sqrt{\frac{2 \times 3.33}{4}}} = -2.5$$

Since the critical two-tailed 5% value for t is ± 2.447, these two means appear to be from different populations. Note that this result is different from the one from the (ambiguous) Tukey test of the same data in Section 11.3.3, which did not separate Runcan and Darcy.

For the comparison between Runcan and Alinda:

$$t_6 = \frac{6.00 - 3.00}{\sqrt{\frac{2 \times 3.33}{4}}} = -2.5$$

Here too, since the critical two-tailed 5% value for t is ± 2.447, these two means also appear to be from different populations. Again, note that this result is different from the one from the Tukey test of the same data in Section 11.3.3, which did not separate Runcan and Alinda. The a priori test has more power.

This example particularly illustrates the value of planned comparisons. Both the Darcy and Alinda sandstones appear to be distinct from the Runcan, at least on the basis of their apatite content. This conclusion is different from the one made on the basis of the less powerful Tukey test (Section 11.3.3) which could not separate the intermediate percentage for Runcan from the higher percentage of Darcy or the lower one of Alinda (and thus included a Type 2 error).

Importantly, you should only make a planned comparison if you really do have a plausible hypothesis to justify this procedure. It is not appropriate or ethical to examine the results of a Tukey test and say, in hindsight, "These two means are almost significantly different, so I will run a t test on them because it is likely to be more powerful."

11.6 Questions

(1) A petroleum scientist was given the following data for the flow rate (in barrels per day on eight days) for three specifically chosen oil wells in

the Timor Sea, and was asked to identify the one which was producing significantly more oil. (a) Does a single-factor ANOVA give a significant result? (b) Is a posteriori testing needed? If so, which well is yielding the most oil and is it significantly different to the other two?

RVB1	RVB2	RVB3
157	183	149
184	174	193
143	182	129
135	199	146
163	183	126
103	168	132
152	193	143
129	162	154

(2) In relation to the data in Question 1, wells RVB1 and RVB3 are only 1 km apart and the oil company was particularly interested in whether the two wells were yielding different amounts. Use a t test to make a planned comparison between wells RVB1 and RVB3 only. Is the result significant?

12 | Two-factor analysis of variance

12.1   Introduction

A single-factor ANOVA gives the probability that two or more sample means have come from populations with the same mean (Chapter 10), but can only be used to analyze univariate data from samples exposed to different levels or aspects of only **one factor**. For example, it could be used to compare the diffusivity of hydrogen through olivine (the variable) at two or more temperatures (the factor), the percentage of feldspar crystals (the variable) in successive layers of an intrusive complex (the factor), or the salinity of seawater (the variable) from several different depths (the factor).

Often, however, scientists obtain univariate data in relation to **more than one factor**. Examples of two-factor experiments are the phase equilibrium of aluminosilicates (Al_2SiO_5) at several combinations of temperature **and** pressure, the growth of crystals as a function of magmatic H_2O **and** cooling rate, or the likelihood of snow as a function of varying humidity levels **and** temperatures.

It would be very useful to have an analysis that gave separate F ratios (and the probability that the treatment means had come from populations with the same mean) **for each of the two factors**. That is what two-factor ANOVA does.

12.1.1   Why do an experiment with more than one factor?

Experiments that simultaneously include the effects of more than one factor on a particular variable may be far more revealing than looking at each factor separately because you may detect certain combinations of factors that have a **synergistic effect**. Also, by examining several factors at once, there may be significant savings in time and resources compared to doing a series of separate experiments and separate analyses.

Here is an example of the advantage of a two-factor experiment. It also illustrates a synergistic effect – what statisticians call **interaction** – which occurs when the effect of one factor varies across the levels of the other.

Research on establishing the importance of the many possible causes of global warming relies on understanding the evolution of greenhouse gasses in the Earth's atmosphere. These originate largely from volcanic eruptions, which release gasses such as CO_2, H_2SO_4, HCl and HF from the interior of the planet. If these build up in the atmosphere, they can increase the amount of solar radiation being absorbed, causing temperatures to increase. We are fortunate that on Earth these gasses react with liquid water to form minerals such as $CaCO_4$ (calcite, or limestone in rock form) and apatite ($CaSO_4$), so end up getting stored in geological deposits, mostly in the ocean, instead of heating up our atmosphere. Incidentally, the planet Venus was not so lucky – it was too close to the Sun for liquid water to be stable on the surface, so all its greenhouse gasses ended up in its atmosphere, and the surface temperature there is an inhospitable 460 °C.

A climatologist investigating paleoclimates on Earth decided to examine the effects of temperature and humidity on the rate of calcite formation, in an attempt to help predict the storage of carbon dioxide in calcite in response to global warming at sub-tropical latitudes. They designed an experiment to examine the amount of calcite precipitation from seawater as a function of both temperature and humidity (see Ufnar *et al.*, 2008, for a related example). Identical beakers of carbonate-rich seawater solutions were placed in six combinations of three temperatures (20, 30 and 40 °C) and two humidity levels (33 and 66%). There were four beakers in each treatment, so 24 were used altogether.

This type of design, where there is a treatment for every combination of the levels of each factor used, is called a "fully orthogonal" design or an "orthogonal" design (Table 12.1). If one of the treatments was not included

Table 12.1 Example of an orthogonal two-factor design. There are three levels of Factor A (temperature) and two levels of Factor B (humidity) with four experimental units (beakers) in each of the six possible combinations of the 3 × 2 treatment levels.

Humidity (%)	Temperature (°C)		
	20	30	40
33	4 beakers	4 beakers	4 beakers
66	4 beakers	4 beakers	4 beakers

(for example the combination of 33% humidity with 20 °C), then the design would not be orthogonal.

The results of the experiment can be displayed as a graph of the means for each of the six combinations (which are often called **cell means**), with temperature on the X axis, the weight of calcite precipitate on the Y axis, and lines joining the three means within each of the two levels of humidity. (If you wanted you could show humidity on the X axis and have lines joining each of the three temperatures, but it is easier to visualize when the greatest number of treatment levels are on the X axis.)

Figure 12.1(a) shows a set of cell means where there is **no interaction** – the change in humidity from 33 to 66% (or from 66 to 33%) has the same effect on calcite precipitation at each temperature (in all cases an increase in humidity increases calcite by about the same amount). Similarly, the effect of an increase in temperature from 20 °C through to 40 °C (or vice versa) is the same at each humidity.

In contrast, Figure 12.1(b) shows **interaction.** A change in humidity from 33 to 66% does not have the same effect on calcite precipitation at each of the three temperatures, and a change in temperature from 20 °C through to 40 °C does not have has the same effect on calcite precipitation at each humidity.

That is all interaction is. When there is a complete lack of interaction (e.g. Figure 12.1(a)) the lines joining the treatment means always run exactly parallel to each other (even though both lines move up, they move up in parallel). In contrast, when there is interaction (e.g. Figure 12.1(b)) the lines are not always parallel. As the amount of interaction increases, the lines become less and less parallel and eventually the amount of interaction may reach a point where it is considered significant.

Interaction between two or more factors is often of great interest to earth scientists. It may be very helpful to know that a response to one factor is not uniform across the range of a second factor, or that it **is** uniform! For example, if you found that the rate of calcite precipitation is unexpectedly high only when temperature is high and humidity is low (Figure 12.1(b)), this synergistic effect would be an important component of models predicting carbon dioxide storage in relation to global warming.

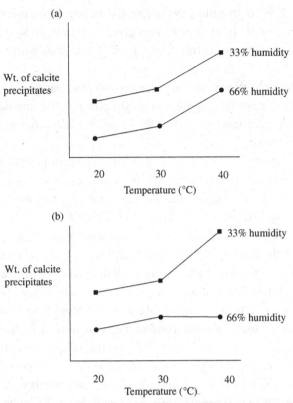

Figure 12.1 Interaction in a two-factor experiment. (a) No interaction between the two factors temperature and humidity on calcite precipitation. A change in humidity from 33 to 66% has the same effect on the amount of calcite precipitation at each of the three temperatures, and a change in temperature from 20 °C through to 40 °C has the same effect at each humidity. (b) An interaction between temperature and humidity on calcite precipitation. A change in humidity from 33 to 66% does not have the same effect on calcite precipitation at each of the three temperatures, and a change in temperature from 20 °C through to 40 °C does not have the same effect on calcite precipitation at each humidity.

12.2   What does a two-factor ANOVA do?

Here you need to remember that a single-factor ANOVA partitions the total variation into two components – the variation among groups (treatment + error) and the variation within groups (error), and examines whether there is a significant effect of treatment by dividing the among groups mean square by the within groups mean square. This gives an F ratio and probability that all the treatment means have come from populations with the same mean.

A two-factor ANOVA works in a similar way, but partitions the total variation within a set of data into **four** components: the among group variation due to (a) Factor A + error, (b) Factor B + error, (c) interaction + error and (d) error.

The way the analysis works is a straightforward extension of the concept developed to explain single-factor ANOVA, and can also be explained pictorially. For this we will use the simplest case of a two-factor design with two levels only of each factor, both of which are fixed.

First, here are some examples of the types of outcomes you might get from a two-factor experiment. We mentioned in Chapter 1 that colored gemstones are often treated with some combination of ^{60}Co irradiation and heat to improve their appearance and commercial value. To better understand the effects of these variables, a gemologist undertook an experiment using four different combinations of heat and irradiation, with four crystals in each treatment. Each of the 16 experimental units was cut from a large amethyst crystal, so the starting composition and appearance of each crystal were exactly the same. The crystals were kept at four combinations of two temperatures and two different ^{60}Co doses: after two months the samples were removed and their opacity (degree of transparency) was examined.

Several different outcomes are shown in Figure 12.2. The opacity within each treatment combination with the 10 kGy radiation dose is indicated by ●, while the treatments exposed to the 100 kGy radiation dose are indicated by ■.

12.3   How does a two-factor ANOVA analyze these data?

This explanation assumes you are familiar with the one already given for a single-factor ANOVA in Chapter 10. We are using a two-factor experiment with two levels of each factor, giving four treatment combinations, each of which contains four replicates. The design is summarized in Table 12.2. Both factors are fixed – the researcher is only interested in these specific temperatures and radiation doses.

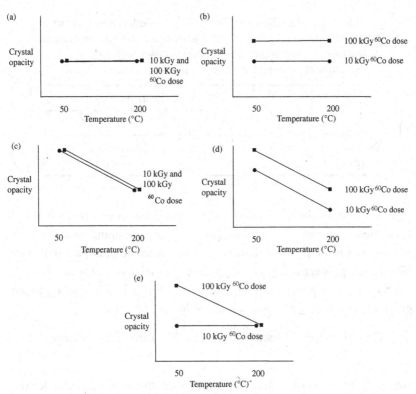

Figure 12.2 Some of the possible outcomes of an orthogonal two-factor experiment. Only the means for each treatment combination are shown. (a) No effect of temperature or radiation dose and no interaction. All treatment means are the same and the lines joining the means within each radiation dose are also the same. (b) An effect of radiation dose, but no effect of temperature and no interaction. The two treatment means for 100 kGy radiation dose are consistently higher than the two for 10 kGy radiation dose. (c) An effect of temperature but no effect of radiation dose and no interaction. The two treatment means for 50 °C are consistently greater than the two for 200 °C. (d) An effect of temperature and radiation dose but no interaction. All treatment means are different, but the change in opacity in relation to a change in radiation dose from 10 to 100 kGy is the same at each temperature and vice versa. (e) An effect of temperature and radiation dose and some interaction. The change in opacity between 10 and 100 kGy dose is not the same at each temperature and vice versa. Note that all lines joining the treatments within the same radiation dose are parallel except in example (e), where there is some interaction between temperature and radiation dose.

Table 12.2 The orthogonal design used to explain how two-factor ANOVA works in Figures 12.3 to 12.6. There are four combinations of the two temperatures and two radiation doses, with four experimental units (in this example, amethyst crystals) in each.

	Temperature (°C)	
^{60}Co dose (kGy)	50	200
10	4 crystals	4 crystals
100	4 crystals	4 crystals

To start, think about the opacity of the amethyst crystals that will result from these treatments. It will be displaced from the grand mean by four sources of variation – that associated with **Factor A plus Factor B plus interaction plus error**. This is called the **total variation** in the experiment. Put formally, the position on the Y axis of each replicate in relation to the grand mean will be determined by the following formula:

$$\text{Crystal opacity} = \text{Factor A} + \text{Factor B} + \text{interaction} + \text{error}$$

$$(12.1)$$

Here you may wish to contrast this with the much simpler equation for the total variation within a single-factor experiment from Chapter 10:

$$\delta^{18}\text{O of tourmaline} = \text{treatment} + \text{error} \quad (12.2 \text{ copied from } 10.1)$$

Just as in the single-factor ANOVA, the variation within a two-factor experiment can be partitioned into several additive components. These are shown in Figures 12.3 to 12.6.

First, the final opacity of each crystal will be displaced from its respective cell mean by error only. This is estimated in just the same way as for a single-factor ANOVA and also called the **within group variation or error** (Figure 12.3). The distances between each replicate and its cell mean are squared and added together to give the within group (error) sum of squares. The sum of squares is divided by the appropriate degrees of freedom (here there are $3 + 3 + 3 + 3 = 12$) to give the within group (error) mean square.

Second, each replicate will be displaced from the grand mean by **all sources of variation in the experiment** – the effect of Factor A plus Factor B plus interaction plus error. This is called the **total variation** in

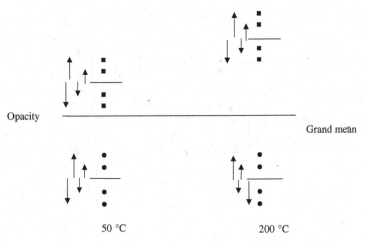

Figure 12.3 The estimation of within group (error) variation in the experiment on gemstone opacity when amethyst crystals are exposed to four different combinations of temperature and radiation dose. Each crystal is shown as a symbol: ● = crystals at 10 kGy ^{60}Co dose, ■ = crystals at 100 kGy ^{60}Co dose. Horizontal lines indicate the grand mean and each cell mean. The displacement of each replicate from its cell mean (arrows) will be caused by error only.

the experiment. In Figure 12.4 the distance displaced is shown for all replicates. These distances can be squared and added together to give the total sum of squares for the experiment. (Again, this is the same as the procedure for a single factor ANOVA.)

So far, this is the same procedure used to calculate the within group (error) variance and total variance for a single-factor ANOVA.

(a) **The within group variance** (Figure 12.3) which is due to error only can be calculated from the dispersion of the points around each of their respective cell means.

(b) **The total variance** (Figure 12.4) will estimate the total variation in the experiment (the within group (error) variance plus Factor A, Factor B, plus interaction) and can be calculated from the dispersion of all the points around the grand mean.

At this stage you still need **separate** effects for Factor A (temperature + error), Factor B (dose + error) and A × B (interaction + error).

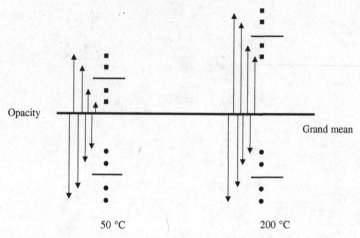

Figure 12.4 The total variation within the experiment on gemstone opacity. Each crystal is shown as a symbol: ● = crystals at 10 kGy ^{60}Co dose, ■ = crystals at 100 kGy ^{60}Co dose. The heavy horizontal line indicates the grand mean. The four shorter horizontal lines indicate each cell mean. The displacement of each replicate from the grand mean (arrows) will be caused by the total variation within the experiment.

12.4 How does a two-factor ANOVA separate out the effects of each factor and interaction?

Two-factor ANOVA separates out the effects of each factor and interaction in a very elegant way. After having done the preliminary calculations in Figures 12.3 to 12.4, the data are only considered in relation to each of the two factors. This is done by first ignoring the different levels within Factor B and considering the data only in relation to Factor A (temperature), after which the same is done for Factor B (radiation dose). These procedures are shown in Figures 12.5 and 12.6 and allow you to calculate separate sums of squares for temperature + error and also dose + error. They are called the **simple main effects** because they examine each factor in isolation from the other.

First, the levels of radiation dose are ignored and the data treated as though they are for a single-factor experiment on temperature only. Here, therefore, you will have eight replicates within each of the two levels of temperature and you can calculate a mean for each group. These new means, calculated from all eight replicates within each treatment, **will only be displaced from the**

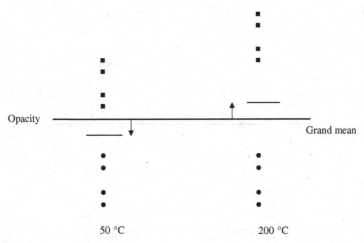

Figure 12.5 The effect of Factor A (temperature + error) only on the opacity of amethyst crystals. Each crystal is represented as a symbol: ● = crystals at 10 kGy ^{60}Co dose, ■ = crystals at 100 kGy ^{60}Co dose. These data have been pooled for each temperature, ignoring radiation dose, thereby generating two new treatment means, shown by the horizontal lines. The displacement of each treatment mean from the grand mean is an estimate of the average effect of temperature plus error. The sum of squares is the sum of the square of each displacement, which is then multiplied by the number of replicates in that treatment. The mean square is the sum of squares divided by $n - 1$ degrees of freedom where n is the number of temperature treatments (here $n = 2$).

grand mean by the average effect of temperature plus error. Therefore, the displacement of the treatment means from the grand mean can be used to calculate the sum of squares and mean square for Factor A (temperature + error) only (Figure 12.5) just as in a single-factor ANOVA.

Second, the levels of temperature are ignored and the data are treated as though they are for a single-factor experiment on radiation dose only. Here too, you will have eight replicates within each of the two levels of radiation dose and you can calculate a mean for each of the two groups. These new means, calculated from all eight replicates within each treatment, **will only be displaced from the grand mean by the average effect of dose plus error** (Figure 12.6). Therefore, the displacement of the treatment means from the grand mean can be used to calculate the sum of squares and mean square for Factor B (radiation dose + error) only just as in a single-factor ANOVA.

At this stage you have sums of squares for the following:

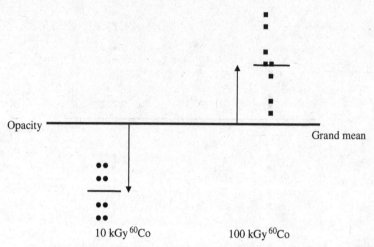

Figure 12.6 The effect of Factor B (radiation dose + error) only on the opacity of amethyst crystals, represented here as symbols: ● = crystals at 10 kGy ^{60}Co dose, ■ = crystals at 100 kGy ^{60}Co dose. These data have been pooled for each radiation dose, ignoring temperature, thereby generating two different treatment means, shown by the horizontal lines. The displacement of each treatment mean from the grand mean is the average effect of dosage for the number of replicates in that treatment. The sum of squares for the effect of radiation dose is the sum of the square of each displacement, which is then multiplied by the number of replicates in that treatment. The mean square is the sum of squares divided by $n - 1$ degrees of freedom where n is the number of radiation treatments (here $n = 2$).

(a) **The total variation in the experiment** (the combined effects of Factor A, Factor B, A × B and error) (Figure 12.4)

(b) **The effect of Factor A** (temperature + error) (Figure 12.5)

(c) **The effect of Factor B** (dose + error) (Figure 12.6)

(d) **error** (Figure 12.3)

From this list, the only separate sum of squares you still need is the one for **interaction plus error**. Because the sums of squares are additive and the total variation is the combined effects of all the factors in the ANOVA (Section 10.3.4), you can calculate the sum of squares for interaction by subtraction. This is done by taking away the sums of squares for Factor A, Factor B, and error from the total sum of squares ((a) above minus (b) and (c) and (d)). Now you have the following sums of squares:

- **The total variation in the experiment** (the combined effects of Factor A, Factor B, A × B and error) (Figure 12.4)
- **The effect of Factor A** (temperature + error) (Figure 12.5)

Table 12.3 Variation estimated by each mean square term and the appropriate division to estimate the effect of each factor when Factor A and Factor B are both fixed.

Source of variation	Calculation of F ratio
Factor A	$\dfrac{\text{Mean square for Factor A}}{\text{Error}}$
Factor B	$\dfrac{\text{Mean square for Factor B}}{\text{Error}}$
Interaction (A × B)	$\dfrac{\text{Mean square for interaction}}{\text{Error}}$

- **The effect of Factor B** (radiation dose + error) (Figure 12.6)
- **The effect of interaction** (interaction + error) (by subtraction)
- **Error** (Figure 12.3)

Once you have these, dividing by the appropriate degrees of freedom will give you mean square values, just as for a single factor ANOVA. The effect of each factor can be estimated by dividing the factor mean square by the error mean square to get an F ratio. If the F ratio is significant, the factor is considered to have an effect. The F ratios for the effects of interaction, Factor A and Factor B are summarized in Table 12.3. Most statistical packages will give an analysis of variance summary table that has all of these sums of squares, degrees of freedom, mean square values and F ratios.

12.5 An example of a two-factor analysis of variance

Gemologists sometimes characterize the appearance of treated gemstones on the basis of their thermoluminesence, which is easily measured in a spectrometer. Quartz crystals may be treated with high doses of radiation to induce defects (color centers) that turn colorless quartz into more valuable smoky quartz (brown). The data in Table 12.4 are for the intensity (measured in counts $\times 10^5$) of the 380 nm thermoluminescence peak in quartz crystals treated at three temperatures and three levels of ^{60}Co radiation dose.

As an initial step, you might plot the cell means on a graph similar to Figure 12.2 to see what they look like. Which factors might you expect to be significant? Would you expect a significant interaction? Why?

Next, if you use a statistical package to run a two-factor ANOVA on these data your results will include something similar to Table 12.5 which gives the F ratio and probability for each of the two factors and their interaction.

Table 12.4 Thermoluminesence (measured in counts $\times 10^5$) of 27 quartz crystals kept in nine different combinations of temperature and ^{60}Co radiation dose.

^{60}Co dose (kGy)	Temperature (°C)		
	50 (level 1)	200 (level 2)	400 (level 3)
10 (level 1)	1	5	9
	2	6	10
	3	7	11
100 (level 2)	9	13	17
	10	14	18
	11	15	19
1000 (level 3)	17	21	25
	18	22	26
	19	23	27

Table 12.5 An example of the type of output given by a statistical package for a two-factor ANOVA.

Source of variation	Sum of squares	df	Mean square	F ratio	Significance
Temperature	312.66	2	156.33	156.33	0.000
Radiation	1200.66	2	600.33	600.33	0.000
Temperature* Radiation	1.33	4	0.33	0.33	0.852
Error	18.00	18	1.00		

The interaction term is symbolized by Temperature*Radiation. Note that the F ratios for temperature and radiation dose are significant at $P < 0.001$, but there is no significant interaction ($P = 0.852$). It seems the samples have come from different populations in relation to the levels of temperature and also radiation dose, but there is no interaction between these factors. This result should not be a surprise if you have plotted the six treatment means before doing the analysis.

12.6 Some essential cautions and important complications

There are some essential cautions and important complications associated with two-factor and more complex ANOVAs that you must be aware of.

(1) A significant effect of a factor does not reveal where differences occur if you have examined more than two levels of that factor.
(2) A significant interaction can make the F ratios for Factor A or Factor B misleading.
(3) If one or both of the factors are random, you need to use a different procedure for calculating the F ratios for one or both of Factors A and B.

These three complications are explained below.

12.6.1 A posteriori testing is still needed when there is a significant effect of a fixed factor

First, just as for a single-factor ANOVA, a significant effect does not reveal where differences occur among the levels of that factor. For example, if you did a two-factor ANOVA with four levels of Factor A and six of Factor B, and found a significant effect of Factor A, it will not identify which levels of Factor A appear to come from populations with the same, or different, means. Here, just as for a single-factor analysis, you need to carry out a posteriori testing. This is straightforward if there is no significant interaction.

If the interaction is not significant a posteriori testing can be done for each factor that has a significant effect. This compares the mean values for the pooled data (e.g. Figures 12.5 and 12.6) in just the same way as a single-factor ANOVA (Chapter 11). For example, if you were to use a Tukey test, the formula is the same as the one given in Chapter 11:

$$q = \frac{\overline{X}_A - \overline{X}_B}{\text{SEM}} \tag{12.3}$$

To calculate the standard error of the mean from the ANOVA statistics you use:

$$\text{SEM} = \sqrt{\frac{\text{MS error}}{n}} \tag{12.4}$$

where n is the sample size of each pooled group. If the sample sizes are different, you need to use a slight modification of the formula (which reduces to the one above when n_A is the same size as n_B).

$$\text{SEM} = \sqrt{\frac{\text{MS error}}{2} \times \left(\frac{1}{n_A} + \frac{1}{n_B}\right)} \tag{12.5}$$

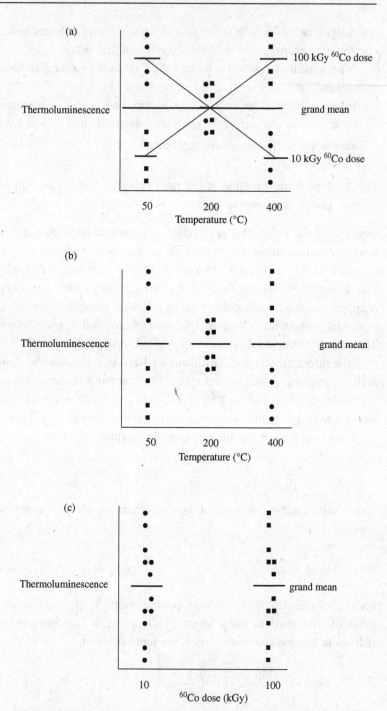

Figure 12.7 An illustration of how interaction can obscure main effects in a two-factor ANOVA. (a) As temperature increases, thermoluminescence

Then you simply calculate the Tukey q statistic for each pair of means and look up the critical value, using the degrees of freedom from the MS within groups (error). If the calculated value is greater than the critical value of q, the hypothesis that the means are from the same population is rejected. The value of q will range from zero when the two sample means are the same to high values as the means become increasingly different. Once again, many statistical packages will do Tukey tests and assign the means to groups that are significantly different to each other.

Just as with a one-factor experiment, a priori planned comparisons can also be made between particular cell means but only if these have been specified beforehand (see Section 11.5).

12.6.2 An interaction can obscure a main effect

The two-factor analysis described in Section 12.5 gave mean squares for the main effects of Factor A (temperature) and Factor B (radiation dose), interaction and also error. The effect of each factor is estimated by dividing the factor mean square by the error mean square.

This is appropriate, but there is a complication. A significant interaction means that the effect of one factor (e.g. radiation dose) is not constant across the levels of the second factor (e.g. temperature). **Therefore, if there is a significant interaction, the conclusion of a non-significant main effect (because of a non-significant F ratio for that factor) may not be correct.**

Here is a rather extreme example which clearly illustrates the problem. Imagine an experiment designed to investigate the effects of two treatments, with three levels of Factor A and two of Factor B. Figure 12.7 shows the results of this experiment. Although there is obviously an effect of temperature and also of radiation dose on thermoluminescence, the response to temperature at 100 kGy ^{60}Co is the opposite of that at 10 kGy ^{60}Co.

Caption for Figure 12.7 (cont.) decreases at 10 kGy dosage, but increases at 100 kGy dosage. (b) When radiation dose is ignored, the cell means for the three levels of temperature only, ignoring dosage, are shown as three short horizontal lines. Note they all lie on the grand mean. The sum of squares for temperature will be zero. (c) When temperature is ignored, the cell means for the two levels of radiation dose only, ignoring temperature, are shown as two short horizontal lines. Note they both lie on the grand mean. The sum of squares for radiation dose will be zero.

When these results are analyzed by a two-factor ANOVA, the total sum of squares will be large because most replicates will be well dispersed from the grand mean (Figure 12.7(a)). There will also be some error because the replicates are dispersed from their cell means (Figure 12.7(a)). But when the ANOVA partitions the sums of squares among the separate factors of temperature and radiation dose, the results are extremely misleading.

First, consider the pooled analysis to assess the effect of temperature. The new cell means for each of the three levels of temperature (ignoring radiation dose) will all lie on the grand mean. Consequently there will be **no overall effect of temperature and the sum of squares for temperature will be zero** (Figure 12.7(b)), even though there is obviously an effect of temperature **within** each level of radiation dose.

Second, consider the pooled analysis to assess the effect of radiation dose. The new cell means for each of the two levels of dosage (ignoring temperature) will also lie on the grand mean, so the sum of squares for radiation dose will also be zero (Figure 12.7(c)) even though there is an effect of dosage within each temperature. The sum of squares for interaction will be realistic and very large.

Therefore, when there is a significant interaction, it is not appropriate to trust the F ratios for the effects of Factors A and B. This caution is particularly important because most statistical packages calculate F ratios for main effects regardless of whether the interaction is significant or not.

The solution to this problem is straightforward. A graph of the cell means such as Figure 12.7(a) is a useful first step, because it will give you a visual indication of the positions of each cell mean. The next step is statistical – you need to look at the effects of each factor across all levels of the second factor using an a posteriori test. This procedure is a little fiddly, but quite easy to do. Here, shown pictorially, is how you can analyze the thermoluminescence example.

First, you compare the two cell means within each of the three levels of temperature (Figure 12.8(a)).

Second, you compare the three cell means within each of the two levels of radiation dose (Figure 12.8(b)).

Here too, for a Tukey test, you simply use the formulae:

$$q = \frac{\overline{X}_A - \overline{X}_B}{\text{SEM}}$$ (12.6 copied from 12.3)

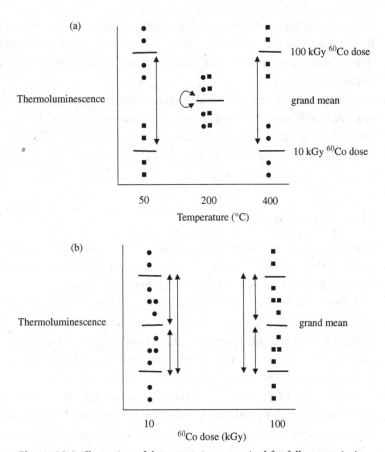

Figure 12.8 Illustration of the comparisons required for full a posteriori testing of a two-factor ANOVA when there is a significant interaction. (a) Double-headed arrows show the means for the two levels of Factor B (radiation dose) within each level of Factor A (temperature) compared as part of full a posteriori testing. (b) Double-headed arrows show the means for the three levels of Factor A (temperature) within each level of Factor B (radiation dose) compared as part of full a posteriori testing.

and

$$\text{SEM} = \sqrt{\frac{\text{MS error}}{n}} \qquad (12.7 \text{ copied from } 12.4)$$

where n is the sample size within each cell. Again, the modification to the formula shown in Equation (12.5) applies if there are different numbers in each cell.

This rather long but extremely important example emphasizes that when there is a significant interaction you need to examine all possible combinations of treatments, and that conclusions from F ratios for main effects may not be realistic.

Most statistical programs will not calculate a posteriori tests for all possible combinations of cell means given above, so it may be necessary for you to do these calculations using a spreadsheet or a calculator. This procedure, and the statistical tables necessary to decide whether each difference is significant, are covered in more advanced texts.

12.6.3 Fixed and random factors

The final complication applies to two-factor and more complex analyses of variance that include random factors. The concept of fixed and random factors was discussed in Section 10.6, but here is a reminder.

A **fixed factor** is one where the treatments (e.g. levels of temperature) have been **specifically chosen**. You are only interested in those particular treatments and the null hypothesis reflects this – for example "There is no difference in crystal opacity after treatment at 50 °C and 200 °C."

A **random factor** is one where the treatments are used as random representatives of the full set of possible treatments within that factor. Therefore, the null hypothesis is more general. Instead of comparing specific temperatures the hypothesis is "There is no difference in crystal opacity at different temperatures." The levels of temperature chosen and used in the experiment are merely **random representatives** of the wider range of temperatures at which opacity or color changes might occur.

For a two-factor ANOVA both factors could be fixed, one could be random and the other fixed, or both could be random.

If a two-factor experiment contains **two fixed factors**, the method for calculating the F ratios for the main effects (Factor A and Factor B) are those given in Table 12.3 and repeated in Table 12.6. The mean square for each factor estimates the effect of that factor plus error, and an F ratio is obtained by dividing the mean square for that factor by the within groups (error) mean square.

If, however, the analysis contains **two random factors**, the sum of squares and mean square for each of the two factors will be **inflated by the inclusion of any additional variation caused by interaction**. Therefore, the variation

Table 12.6 Sources of variation contributing to the mean squares for Factor A, Factor B and interaction when both A and B are fixed, A is fixed and B is random, and both A and B are random.

Source of variation	Both factors fixed	Factor A fixed, Factor B random	Both factors random
Factor A	Factor A + error	Factor A + interaction + error	Factor A + interaction + error
Factor B	Factor B + error	Factor B + error	Factor B + interaction + error
Interaction	Interaction + error	Interaction + error	Interaction + error

estimated by the mean square for each main effect will be the effect of that factor, plus interaction plus error. This is explained pictorially below. Most importantly, to realistically estimate the F ratios for each random factor you need to divide the factor mean squares by the interaction MS (which estimates interaction plus error) rather than the error MS (Table 12.6).

Finally, if the ANOVA has **one fixed and one random factor** it is even more complicated. Most statisticians recommend that if Factor A is fixed and Factor B is random, the F ratio for Factor A is obtained by dividing the Factor A MS by the interaction MS, but the F ratio for Factor B is obtained by dividing the Factor B MS by the error MS (Table 12.6). In all cases the F ratio for interaction is obtained by dividing the interaction MS by the error MS.

Importantly, many statistical packages do **not** give appropriate F ratios when random factors are included in an analysis, so you have to do these calculations yourself by dividing by the appropriate mean squares.

Here is a conceptual pictorial explanation for the different ways of estimating main effects in a two-factor ANOVA depending on whether the other factor is fixed or random. In all cases the fixed factor of interest is Factor A.

Imagine the hypothetical case where the only levels of Factor A and B that exist in the world are A1 and A2, and B1, B2, B3 and B4. As an example, A1 and A2 may be 1wt% and 2wt% H_2O in a magma, where there are four different mineral species crystallizing (B1 to B4). You are interested in the effects of H_2O on the modal abundances of these four minerals.

Figure 12.9(a) shows crystal abundance (by mode, which is a percentage of the total) for all eight possible combinations of Factors A and B. Note that

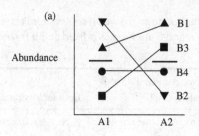

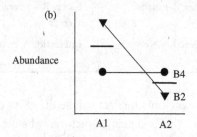

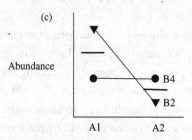

Figure 12.9 A pictorial explanation for the reason why the *F* ratio for a main effect is calculated differently, depending on whether the other factor is fixed or random. Cell means are indicated by symbols and pooled treatment means are indicated by the two heavy horizontal lines. (a) All the possible levels of Factor A and Factor B, together with all possible combinations of these, are shown. Note that there is considerable interaction but overall there is no effect of Factor A (when Factor B is ignored, the pooled treatment means for A1 and A2 are identical). (b) When Factor B is fixed and only a subset of B is considered (B2 and B4), the interaction will contribute to the difference between the pooled means of A1 and A2, but this variation is a relevant addition within the deliberately restricted levels of each factor being compared. (c) When Factor B is random, the interaction will contribute unrealistic additional variation to the difference between the pooled means of A1 and A2. It will not indicate the true lack of change from A1 to A2 across the entire set of the levels of B and therefore needs to be excluded.

there is no effect of Factor A when averaged over all possible levels of Factor B because the means for each of the levels A1 and A2, ignoring the separate levels of Factor B, are all the same. Nevertheless, there is considerable interaction between the two factors.

Both factors fixed and an interaction

First, consider the case where **both factors are fixed**, and you are **only interested in the four combinations of A1 and A2 with B2 and B4**. Because both factors are fixed, you are **not** interested in whether any differences in modal abundance between A1 and A2 within this very restricted comparison also reflect those averaged over all possible levels of Factor B.

The comparisons between A1, A2 and B2, B4 are shown in Figure 12.9(b). Cell means have been copied from the appropriate part of Figure 12.9(a). **Although the means of treatments A1 and A2 (ignoring B) are affected by the interaction,** you are only interested in treatment A1 compared to A2 **within the two fixed levels of B2 and B4.** Therefore, to get a realistic effect of Factor A within this limited and fixed comparison, the variation due to the interaction is a necessary additional component of Factor A and you calculate the F ratio for Factor A by dividing its treatment mean square by error only.

Factor A fixed, Factor B random and an interaction

Second, consider the case where **Factor A is fixed and Factor B is random.** You are interested in the comparison between A1 and A2 across **all possible levels of B**, from which B2 and B4 have been chosen as random representatives. The results of the experiment on the combinations of A1, A2 and B2, B4 are shown in Figure 12.9(c). Here too, the pooled means of treatments A1 and A2 (ignoring B) are affected by the interaction, but the difference within the experiment **does not reflect the lack of change between A1 and A2 averaged over all possible levels of Factor B** in Figure 12.9(a). Therefore, because the interaction has contributed additional variation to the sum of squares and mean square for Factor A it is appropriate to exclude it by dividing the Factor A mean square by the interaction + error mean square to get a more realistic effect of Factor A averaged over all four possible levels of B.

For any two-factor ANOVA, the effect of a particular factor (e.g. Factor A) is estimated by dividing by the mean square for error only if the other factor is fixed, but by the mean square for interaction (i.e. interaction + error) if the other factor is random. Therefore, if both factors are random, you divide the mean squares of both by the interaction mean square.

Finally, although we have specified the procedure for obtaining realistic F ratios when one or both factors are random, there is still some disagreement about this. Some authors recommend dividing the mean square for Factor A and also Factor B by the mean square for interaction + error when either or both is random. Most importantly, if you have an analysis involving one or more random factors it is important to clearly specify how you calculated the F ratios for each factor.

12.7   Unbalanced designs

The cautions about unbalanced designs (when the sample size is not the same in each treatment) in relation to one-factor ANOVA also apply to more complex models. Whenever possible, you should try to ensure that samples sizes are equal in each treatment combination, especially when sample sizes are relatively small, because they may not give good estimates of cell means and result in misleading conclusions.

12.8   More complex designs

Once you understand the concept of single-factor and two-factor analyses of variance, extension to three or more factors and other designs is relatively easy.

A two-factor ANOVA breaks the analysis down into two main factors (which are each analyzed like a single-factor ANOVA) and generates an interaction term by subtraction.

A three-factor ANOVA does the same thing, but the analysis and ANOVA table are more complex because there are three main factors (Factors A, B and C), plus interaction among all three ($A \times B$, $A \times C$, $B \times C$, $A \times B \times C$), and error. More advanced texts give rules for obtaining the appropriate F ratios with more complex designs, where there can be several combinations of fixed and random factors as well.

If you continue on to use ANOVA a lot you will realize that this chapter is very introductory. There are nested ANOVAs, two-factor ANOVAs without replication, ANOVAs for split plot designs, unbalanced designs and many more. This book does not attempt to cover all of these. Instead, it provides you with a general conceptual view that will help you work with more complex designs. If you have to do complex experiments requiring complicated ANOVA, you will need a good advanced textbook (e.g. Koch and Link, 2002; Davis, 2002; Borradaile, 2003; Gamst, et al., 2008). It may help to talk to a statistician before you design the experiment.

12.9   Questions

(1) Constructing your own data set will help you understand how a two-factor ANOVA works and what the F ratios and probability values for each term mean. Use a simple design with three levels of Factor A and two of Factor B. Assume the ANOVA is Model I. First, make all the cells means identical by using the following data:

Factor A	A1		A2		A3	
Factor B	B1	B2	B1	B2	B1	B2
	1	1	1	1	1	1
	2	2	2	2	2	2
	3	3	3	3	3	3
	4	4	4	4	4	4

(a) Analyze these data with a two-factor, Model 1 ANOVA. What are the F ratios and probabilities for each factor and the interaction? (b) Now, deliberately change the data so you would expect a significant effect of Factor B, no effect of Factor A and no interaction. Rerun the two-factor analysis. What are the F ratios and probabilities for each factor and the interaction? (c) Finally, deliberately change the data so you would expect a significant effect of Factors A and B, but no interaction, and rerun the analysis. What are the F ratios and probabilities for each factor and the interaction? It will help if you start this problem by drawing a rough graph of the cell means like the one in Figure 12.7(a).

(2) In the previous question you examined a simple design with three levels of Factor A and two of Factor B. Change the data in the table given in Question 1 so you would expect a significant effect of Factor A and Factor B as well as a significant interaction. Run the two-factor analysis. (a) What are the F ratios and probabilities for each factor and the interaction? Here too it will help if you draw a rough graph of the cell means like the one in Figure 12.7(a) to visualize the data.

13 | Important assumptions of analysis of variance, transformations and a test for equality of variances

13.1 Introduction

Parametric analysis of variance assumes the data are from normally distributed populations with the same variance and there is independence, both within and among treatments. If these assumptions are not met, an ANOVA may give you an unrealistic F statistic and therefore an unrealistic probability that several sample means are from the same population. Therefore it is important to know how robust ANOVA is to violations of these assumptions and what to do if they are not met, because in some cases it may be possible to transform the data to make variances more homogeneous or give distributions that are better approximations to the normal curve.

This chapter discusses the assumptions of ANOVA, followed by three frequently used transformations. Finally, there are descriptions of two tests for the homogeneity of variances.

13.2 Homogeneity of variances

The first and most important assumption is that the data for each treatment (or treatment combination in the case of two-factor and more complex ANOVA designs) are assumed to have come from populations that have the same variance. Equality of variances is called **homogeneity of variances** or **homoscedasticity**, while unequal variances show **heterogeneity of variances** or **heteroscedasticity**. Nevertheless, statisticians have found that ANOVA is relatively robust in terms of departures from homoscedasticity, and there has been considerable discussion about whether it is necessary to apply tests which assess this before doing an ANOVA, especially because these may be too sensitive when sample sizes are large, or too insensitive when sample sizes are small (e.g. Koch and Link, 2002). Many authors suggest preliminary testing for homoscedasticity is **not** necessary, providing as a very general rule that the ratio of largest variance to the smallest variance does not exceed 4 : 1.

Some cases of heteroscedasticity can be reduced by transforming the data (Section 13.5). Consequently, it is often useful to plot the data or calculate the variance within each treatment, or treatment combination, to see if there is a trend. For example, geological data often show an increase in variance as the mean increases, in which case transforming the data by taking the square root of each value may reduce heteroscedasticity (Section 13.5).

There are several tests designed to assess heteroscedasticity and these have more uses than just checking whether data are suitable for parametric analysis. Sometimes you may be interested in a hypothesis about the **variances** rather than the means of different treatments. For example, you might hypothesize that cooling rate affects the variance of quartz abundance in granite, so you would need to analyze your data with a test that compares variances among different localities. The Levene test for heteroscedasticity is described in Section 13.7.

13.3   Normally distributed data

The second assumption is that the data are from normally distributed populations. Nevertheless, it has been shown that ANOVA is quite robust in terms of minor departures from normality. As previously described in Section 8.7.1, drawing P-P plots can assess normality. You should only be cautious about proceeding with a parametric analysis if a P-P plot shows gross departures from linearity such as sharp kinks.

13.3.1   Skew and outliers

A box-and-whiskers plot (Tukey, 1977) is a way of visually summarizing the distribution of a sample (Figure 13.1) so it can be assessed for **skew** and whether there are values in the data set which are unusually distant (either greater or less) from the mean. These are called **outliers**. Construction of a box-and-whiskers plot is straightforward.

For a sample containing an odd number of values you need to find the median, which is the middle value of the set of data.

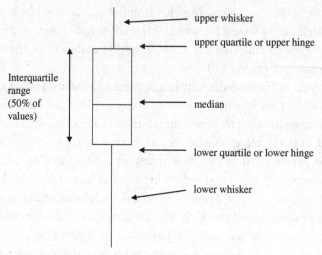

Figure 13.1 The features of a box-and-whiskers plot.

Next, divide the data into two sets, the first of which contains all the values less than the median and the second of which contains all the values more than the median. Include the median in each set.

Then, find the median of each of the lower and upper set. These new medians are called the **lower quartile** and **upper quartile**, which are used to draw the upper and lower limits (which are also called the hinges) of the box. The distance between these quartiles or hinges is the **interquartile range**. Twenty-five percent of the values in the sample will be larger than the upper quartile, fifty percent will lie between the two quartiles and twenty-five percent will be smaller than the lower quartile.

Finally, you need to add the whiskers to the box. Each whisker can extend outwards for a maximum distance of 1.5 times the interquartile range from each end of the box, but is only drawn to the **maximum value within that range**.

This will give you a plot with a box running from the lower to upper quartiles and whiskers extending out from each end of the rectangular box (Figure 13.1).

For a data set with an even number of values the procedure is almost the same except that after finding the median you divide the data into two sets, the first of which contains all the values less than the median and the second of which contains all the values more than the median.

13.3.2   A worked example of a box-and-whiskers plot

This example uses a sample with an odd number of values ($n = 9$): 1, 3, 4, 6, 7, 9, 10, 12, 25. The median of this sample is 7, so it is divided into two groups where the lower group contains 1, 3, 4, 6 and 7, while the upper group contains 7, 9, 10, 12 and 25. The median of the lower group is 4, which becomes the lower quartile. The median of the upper group is 10, which becomes the upper quartile. These are the limits of the ends of the box (called the hinges).

The interquartile range is $10 - 4 = 6$ units. From this you can draw the rectangular box in Figure 13.2(a). The **maximum** potential length of each whisker is 1.5 times the interquartile range and thus $1.5 \times 6 = 9$. This is shown in Figure 13.2(b). Each whisker can extend out a maximum of 9 units from its hinge. Because each whisker is only drawn **to the most extreme value within its potential range**, the lower whisker will only extend down to 1, while the upper will only extend up to 12. The outlier of 25, indicated by an asterisk, lies outside the range of the box and its whiskers (Figure 13.2(c)).

The **shape** of the box-and-whiskers plot indicates whether the distribution is skewed. If the distribution of the data is symmetrical about the mean the box-and-whiskers plot will have a median equidistant from the hinges, and whiskers that are of similar length. As the distribution becomes increasingly skewed the median will become less equidistant from the hinges and the whiskers will have different lengths (Figure 13.3).

Any values outside the range of the whiskers are called **outliers** and should be scrutinized carefully. In some cases outliers are obvious mistakes caused by incorrect data entry or recording, faulty equipment or inappropriate methodology (e.g. a daily temperature of $-80\,^{\circ}$C or a negative radiometric age date) in which case they can justifiably be deleted. When outliers appear to be real, they are of great interest because they may indicate that something unusual is occurring, especially if they are present in some samples or treatments and not others. Importantly, however, when there are outliers you should be cautious about using a parametric test. One or two extreme values can greatly affect the variance of a sample because the formula for the variance uses the square of the difference between each value and the mean, so the

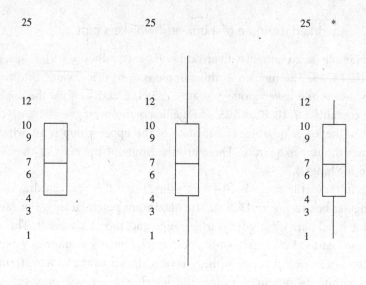

| 25 | 25 | 25   * |

(a) The box

(b) The maximum potential range of each whisker

(c) The actual range of each whisker

Figure 13.2 The three steps in drawing a box-and-whiskers plot, using the data in Section 13.3.2. (a) Drawing the box. (b) Establishing the maximum potential length of each whisker. (c) Drawing the actual length of each whisker.

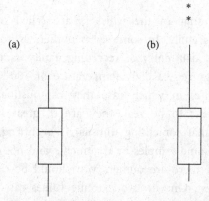

(a)

(b)

Figure 13.3 Examples of box-and-whiskers plots for (a) normally distributed data and (b) data with a gross positive skew. Outliers are shown as asterisks.

assumption of equal variances among treatments or samples can be easily violated.

13.4 Independence

Finally, the data must be independent of each other, both within and among groups. This important assumption needs very little explanation because it is really just a matter of good experimental design. For example, you need to ensure each sampling or experimental unit within each treatment, or combination of treatments for more complex designs, is chosen independently and all possible units within the population have an equal likelihood of being selected.

13.5 Transformations

Transformations are a way of reducing heteroscedasticity or making data more closely resemble a normal distribution. There are many transformations available, and three commonly used ones are described below. Most spreadsheet and statistical packages include a large choice of transformations.

13.5.1 The square root transformation

If the variance of the data increases as the mean increases, a square root transformation will make these data more homosecdastic. There is an example in Table 13.1.

13.5.2 The logarithmic transformation

If the data show a gross positive skew, a logarithmic transformation will give a distribution that better approximates one that is normal. Many types of naturally occurring phenomena have logarithmic distributions, such as crystal size populations in magmatic rocks, river basin sizes, island sizes and star brightnesses.

In cases where the data set includes any values of zero you need to use the logarithm of $X + 1$ because the logarithm of zero is $-\infty$. Many types of geological data, especially those based on biological processes

Table 13.1 An example of the effect of a square root transformation on data where the variance increases as the mean increases. Data are given for the porosity of sandstones at three different drill sites with oil potential. The original data show gross heteroscedasticity among groups in that the largest variance is 10.92 and the smallest is 0.67, giving a ratio of largest to smallest of 16.38 : 1. A square root transformation reduces this ratio to 2.05 : 1.

Clear Lake		Webster		Seabrook	
Original	Square root	Original	Square root	Original	Square root
19	4.36	7	2.65	3	1.73
15	3.87	5	2.24	2	1.41
14	3.74	4	2.00	2	1.41
11	3.32	3	1.73	1	1.00
s^2 10.92	0.18	2.92	0.15	0.67	0.09

(e.g. dimension of fossils) show a positive skew, and Figure 13.4 shows the effect of a logarithmic transformation on a positively skewed distribution.

13.5.3 The arc-sine transformation

The arc-sine transformation can be useful for data that are percentages. Because percentage data have an absolute minimum of 0% and an absolute maximum of 100%, any distribution with a mean close to either of these extremes is unlikely to have a normal distribution because it will cease at these values (Figure 13.5). An arc-sine transformation will give the distribution a far more normal shape.

13.6   Are transformations legitimate?

Here you may be thinking that transforming data to make them more suitable for parametric statistical analysis sounds like cheating or altering the data to get the result you want.

First, however, transformations are applied to the entire data set, so each value is treated in the same way.

Second, there is no scientific necessity to use the linear base ten scale that we are so familiar with. Many geological relationships between two variables

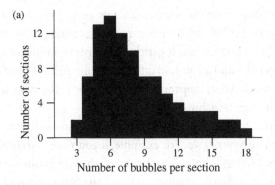

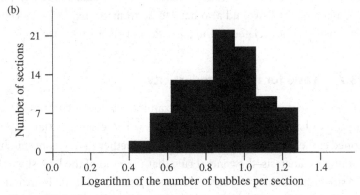

Figure 13.4 The effect of logarithmic transformation on data for the number of bubbles measured on $1\,cm^2$ thin sections of volcanic rock. The X axis shows the number of bubbles per section and the Y axis shows the number of thin sections measured that contained each number of bubbles. (a) The data show a pronounced positive skew before transformation. (b) After transformation to the $\log_{10}$. Note that the distribution in (b) is far more symmetrical than (a).

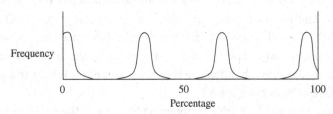

Figure 13.5 Restriction of the normal distribution for percentage data when the mean is close to zero or 100%.

(e.g. earthquake magnitude on the Richter scale vs. energy released, crystal population density vs. crystal size) are logarithms, squares or cubes. The apparently linear pH scale is actually logarithmic – a pH of 4 indicates a ten-fold difference from pH 5 and a 100-fold difference from pH 6. Therefore, in many cases it is actually more appropriate to transform the data so they reflect the underlying relationship.

Importantly, if you transform a set of data, you also need to transform your null and alternate hypotheses. For example, if you were to hypothesize that "the steepness of river banks is not related to river basin size" but carried out a logarithmic transformation on your data before analysis, your original hypothesis would also have to be transformed to "the steepness of river banks is not related to the logarithm of river basin size".

13.7　Tests for heteroscedasticity

There are several tests designed to examine whether two or more samples appear to have come from populations with the same variance. As mentioned earlier, if you are only interested in whether the data are suitable for a parametric analysis, the general rule that the ratio of the largest variance to the smallest should not exceed 4 : 1 can be used. If this ratio is greater, it may be useful to examine the data and see where the differences occur because it may be possible to transform the data so that a parametric analysis can be done.

If, instead, you are interested in testing an hypothesis about the **variance** of two or more samples, you can use the Levene test, which also gives an F ratio. Remember, however, that a significant result for the Levene test may not mean the data are unsuitable for analysis by ANOVA, which is quite robust to heteroscedasticity.

Levene's original test calculates the **absolute difference** between each replicate and its treatment mean and then does a one-factor ANOVA on these differences. The absolute difference is the difference between any two numbers expressed as a positive value. (For example, the difference between 6 and 3 is –3, while the difference between 3 and 6 is +3, but the **absolute difference** in both cases is +3.)

Figures 13.6 and 13.7 are a pictorial explanation of the Levene test. Two cases are shown, using the data on apatite abundance in sandstones that were first described in Section 11.3.3.

(a)

(b)

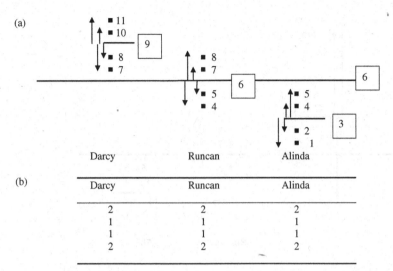

Figure 13.6 The Levene test examines whether two or more variances are likely to have come from the same population by doing a single-factor ANOVA on the absolute differences between the replicates and their treatment means or cell means. (a) Arrows show the difference between each replicate and its treatment mean. Note that some differences are positive and some are negative. (b) The absolute differences are listed under each treatment. Every value of the absolute difference between each replicate and its sample mean will be positive. In this case the means of the absolute differences are the same for each treatment and a single-factor ANOVA comparing these will not be significant, thereby indicating the variances are homoscedastic.

First, if the variances within all treatments are similar, then the set of absolute differences between the replicates and their sample means will also be similar for each treatment. For example, Figure 13.6 shows the absolute differences for three samples that all have the same variance. Note that the means of the absolute differences in 13.6(b) are the same, even though the treatment means in 13.6(a) are not. A one-factor ANOVA comparing the means of the absolute differences will not be significant.

Second, if the variances differ among treatments (Figure 13.7(a)) then so will the values of the absolute differences (Figure 13.7(b)). Note that the set of **absolute differences** for Darcy has a mean that is much larger than the other two. A single-factor ANOVA comparing these means is

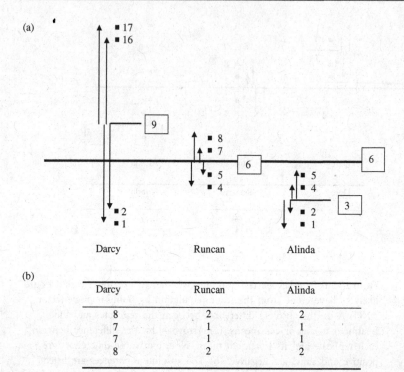

(a)

Darcy   Runcan   Alinda

(b)

Darcy	Runcan	Alinda
8	2	2
7	1	1
7	1	1
8	2	2

Figure 13.7 An example of the Levene test where there is heteroscedasticity. (a) Arrows show the difference between each replicate and its sample mean. (b) The absolute differences between each replicate and its sample mean are listed under each treatment. Because the absolute differences for Darcy are much greater than the other two treatments, a single-factor ANOVA comparing the means of the values in (b) will show the variances are significantly heteroscedastic.

likely to be significant. The Levene test is available in most statistical packages.

13.8   Questions

(1) Why can a transformation be useful when analyzing data with parametric tests, especially ANOVA?

(2) You have been given the following set of data for apatite abundance in sandstones at four outcrops. (a) Do the data require transformation before running a single-factor ANOVA comparing the four locations? What transformation would you recommend?

Darlinghurst	Glebe	Newtown	Kiama
15	4	2	7
9	2	4	9
12	3	3	6
18	1	5	10

14 | Two-factor analysis of variance without replication, and nested analysis of variance

14.1 Introduction

This chapter describes two slightly more complex ANOVA models often used by earth scientists, but an understanding of these is **not** essential if you are reading this book as an introduction to geostatistics. If, however, you need to use more complex analyses then the explanations given here for two-factor ANOVA without replication and nested ANOVA are straightforward extensions of the pictorial descriptions in Chapters 10 and 12 and will help with many of the ANOVA models used to analyze more complex designs.

14.2 Two-factor ANOVA without replication

This is a special case of the two-factor ANOVA described in Chapter 12. Sometimes an orthogonal experiment with two independent factors has to be done without replication because there is a shortage of sampling units or the experimental treatments are very expensive. The simplest case of ANOVA without replication is a two-factor design. You cannot do a single-factor ANOVA without replication.

Here is an example of a two-factor design without replication. Oil geologists often drill cores to determine the extent of an oil-rich shale deposit. Time and financial constraints usually mean only one core is taken every kilometer or so away from the suspected location of the deposit. Often oil content also varies with depth, and these data are extremely important for evaluation of the oil field, but you can only afford to log a few sections (at specific depths) of each core. Therefore, a design where only one core is taken every kilometer is orthogonal but unreplicated (Figure 14.1) and care must be taken to avoid confusing variation within

Table 14.1 The length of feldspar phenocrysts (in μm) for a crystallization experiment with nine different combinations of pressure and temperature. All replicates were initially identical compositions. Only one replicate is available for each combination of the two treatments.

	Pressure (MPa)		
Temperature (°C)	100	50	0.1
700	81	76	79
800	45	46	45
900	28	27	27

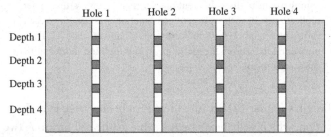

Figure 14.1 When only one core is taken at each location and only one log (sampling unit) is available at each depth (dark squares) the sampling design is orthogonal but unreplicated because only one datum is available for each combination of drill hole and depth.

each core with any variation among drill holes at different locations. An analogous situation is also frequently encountered in environmental geochemistry, where cores are often drilled at different distances from a hazardous waste site to assess the extent (and depth) of contamination.

Another example is given in Table 14.1, which shows data for an experiment to test the effects of pressure and temperature on the growth of feldspar crystals. Pressures of 50 and 100 MPa were being evaluated, together with a control treatment (0.1 MPa, which is ambient room pressure), for their effect in combination with three different temperatures. The experiments are difficult and time-consuming, so only one replicate at each of these nine combinations of temperature and pressure could be run, giving a two-factor orthogonal design without replication.

This causes a problem. There is no way to directly estimate error from the dispersion of replicates around their respective cell means (as was done for a single-factor ANOVA in Chapter 10 and a two-factor ANOVA with

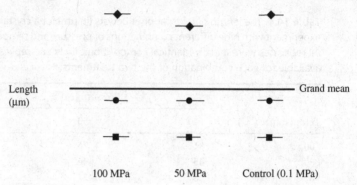

Figure 14.2 Feldspar phenocryst size in crystallization experiments for nine combinations of three different pressures and temperatures. There is only one replicate within each treatment combination. The length of each crystal is shown as a symbol: ♦ = 700 °C, ● = 800 °C and ■ = 900 °C. The heavy horizontal line shows the grand mean and the nine shorter horizontal lines show each mean.

replication in Chapter 12) because there is only one value in each treatment combination, which will always be the same as the cell mean. A two-factor ANOVA without replication uses a different way of estimating error, which has to assume there is no interaction between the factors. Figures 14.2 to 14.5 give a pictorial explanation of how a two-factor ANOVA without replication estimates three sources of variation and uses these to isolate the effects of the two factors. The data in Table 14.1 are graphed in Figure 14.2.

First, the total variation within the experiment is estimated. Each point will be displaced from the grand mean by the effects of Factor A, Factor B, any interaction and error. These distances can be squared and summed to give the sum of squares for the total variation in the experiment, with degrees of freedom that are one less than the number of experimental subjects.

Second, the effect of Factor A is estimated by ignoring Factor B and calculating a new mean for each of the levels within Factor A. The displacement of each treatment mean from the grand mean will be caused by the average effect of Factor A plus error (Figure 14.4). Each of these displacements is squared, multiplied by the number of replicates within each treatment and added together to give the sum of squares for Factor A. The number of degrees of freedom is one less than the number of treatments,

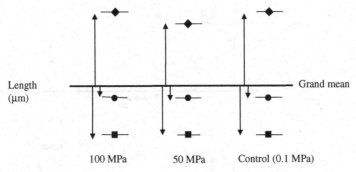

Figure 14.3 The total variation within the feldspar crystallization experiment. The heavy horizontal line indicates the grand mean and the nine shorter horizontal lines indicate each cell mean. The displacement of each point from the grand mean (arrows) will be caused by the total variation within the experiment.

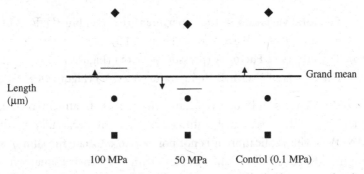

Figure 14.4 Estimation of the effect of Factor A. The displacement of each treatment mean from the grand mean (arrows) will be caused by the effect of Factor A (here pressure) plus error.

and dividing the sum of squares by this value will give the mean square for Factor A.

Finally, the effect of Factor B is estimated by ignoring Factor A and calculating a new mean for each treatment level of Factor B. The displacement of each treatment mean from the grand mean will be caused by the effect of Factor B plus error (Figure 14.5). Here too, the displacements are squared, multiplied by the number of replicates within each treatment, and added together to give the sum of squares for Factor B. The number of degrees of freedom is one less than the number of treatments, and dividing by this value will give the mean square for Factor B.

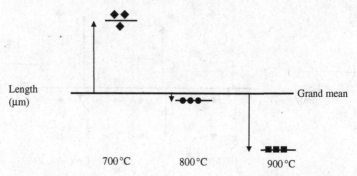

Figure 14.5 Estimation of the effect of Factor B. The displacement of each treatment mean from the grand mean (arrows) will be caused by the effect of Factor B (here temperature) plus error.

At this stage you have estimates for the following sources of variation:

(a) **The total variation in the experiment** (the combined effects of Factor A, Factor B, A × B and error) (Figure 14.3)
(b) **The effects of Factor A** (pressure + error) (Figure 14.4)
(c) **The effects of Factor B** (temperature + error) (Figure 14.5)

Because there is only one replicate within each treatment combination, there is no way to separately estimate error. Therefore, unlike a two-factor ANOVA with replication, it is not possible to estimate the sum of squares for the effect of any interaction by subtracting the sums of squares for Factor A, Factor B and error from the total variation.

Two-factor ANOVA without replication does the next best thing. The sums of squares and degrees of freedom in an ANOVA are **additive** (e.g. in Chapter 10 it was explained how the total sum of squares and total degrees of freedom in a single-factor ANOVA were the sums of those for the sums of squares for Factor A and for error). Therefore, by subtracting the sums of squares for Factor A plus Factor B from the total variation, you are left with the sum of squares for the remaining variation in the experiment, which will include error and any effect of interaction. This sum of squares, which is the only possible estimate of error, is divided by the remaining degrees of freedom to give the best available estimate of the mean square for error. If there is an interaction the mean square will be inflated, but this is unavoidable and undetectable if you do a two-factor ANOVA without replication.

Table 14.2 Results of a two-factor ANOVA without replication on the data in Table 14.1. There is a significant effect of temperature but no significant effect of pressure on the growth of feldspar phenocrysts.

Source of Variation	Sum of Squares	df	Mean square	F	P
P (MPa)	4.222	2	2.111	0.864	0.488
T (°C)	4070.222	2	2035.111	835.545	0.000
Error	9.778	4	2.444		
Total	4084.222	8			

The results of a two-factor ANOVA without replication will include the sums of squares and mean squares for Factor A, Factor B, and error, together with the F ratios and probabilities for Factors A and B. For the example given above the results of the analysis are in Table 14.2.

14.3 A posteriori comparison of means after a two-factor ANOVA without replication

If a two-factor ANOVA without replication shows a significant effect of a fixed treatment factor (e.g. the three temperatures that are specifically compared in Section 14.2), then you are likely to want to know which treatments appear to be from the same or different populations.

The procedure for a posteriori testing is a modification of the formula for a single-factor ANOVA, except that because there is no directly estimated value for error, the MS error for interaction plus error (estimated by subtraction) is used as the best estimate of this. For a Tukey test, each factor is examined separately using the formula:

$$q = \frac{\bar{X}_A - \bar{X}_B}{\text{SEM}}$$ (14.1 copied from 11.1)

with the standard error of the mean estimated from:

$$\sqrt{\frac{\text{MS error}}{n}}$$ (14.2 copied from 11.3)

where the MS error is the one calculated by subtraction in the ANOVA table (see Table 14.2) and n is the number of data within each group (for example, there are three values within each of the three pressures when temperature is ignored and vice versa).

14.4   Randomized blocks

Experiments done in environments where there is **considerable spatial variation** (e.g. a 20-acre mining lease) have to be replicated, but often spatial variation is so great that it may obscure any effect of treatment if replicates were simply assigned at random within that area. One solution is to distribute replicates of each treatment fairly evenly across a landscape. This is often done by setting out a two-dimensional array, with the area subdivided into a series of strips, called **blocks**, with every treatment represented in each. Often only one replicate is available in each block but these data can be analyzed as a two-factor ANOVA without replication, with blocks as a random factor and treatments as a fixed or random factor. This is called a **randomized block design** and gives a way of separating the effects of location and treatment.

Here is an example. Pearls (which mainly consist of calcium carbonate) are usually grown by placing oysters in fine-mesh catch bags attached to chains hanging down at regular intervals from a series of taut horizontal subsurface longlines running parallel to each other (Figure 14.6(a)). An aquacultural scientist hypothesized that removal of marine parasites from the oysters would increase the proportion that produced marketable pearls. Unfortunately, factors including water depth and temperature, wind exposure, light levels, turbidity and tidal currents may vary from chain to chain and longline to longline. If you simply used an experimental design with replicates of each anti-parasite treatment allocated at random to the array you are likely to get a lot of variation among replicates of the same treatment.

For a randomized block design, a set of five parallel longlines (called blocks 1–5) was established. Four bags, each containing 100 oysters, were suspended at regular intervals from every longline and one replicate of the four treatments was assigned at random within each. After six months the number of oysters with pearls was counted in every bag, thus giving only a single value at each point in the array (Figure 14.6(b)).

The results from this design can be analyzed as a two-factor ANOVA without replication, using treatments as the first factor and blocks as the second, thereby subdividing the variation into two components in order to isolate the effect of treatment from any spatial variation.

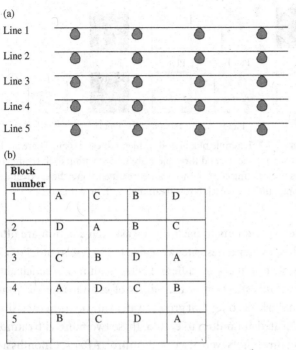

Figure 14.6 A randomized block design. (a) Aerial view of a typical rectangular array used to grow pearl oysters. Five horizontal longlines are run parallel to each other and bags of 100 oysters (filled symbols) are suspended from four chains attached at regular intervals to the longlines. (b) Representation of the array shown in (a) as a 5 block × 4 column grid. One replicate of each treatment (A–D) is assigned at random to a bag within each (longline) block.

14.5 Nested ANOVA as a special case of a single-factor ANOVA

An experimental design that compares the means of two or more levels of the same factor (e.g. different levels of salinity or different compositions) can be analyzed by a single-factor ANOVA, as described in Chapter 10. Sometimes, however, researchers do an experiment with two or more levels of a particular factor, but also have **two or more subgroups nested within each level**. Here is an example from the contaminated landfill first mentioned in Chapter 2. The idea is that mixing the heavy-metal-contaminated soils with apatite group minerals will make the metals bind with the apatite instead of the soil, thereby preventing them from leaching into groundwater.

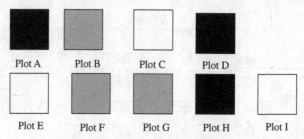

Figure 14.7 Example of a nested or hierarchical design. There are three different treatments, and three plots are nested within each treatment. Open boxes indicate the control (no apatite treatment), grey boxes treatment 1 (chlorapatite) and black boxes treatment 2 (fluorapatite).

Large-scale experiments in hazardous waste remediation are often constrained by the number of test sites available to the researcher. For example, only nine large test plots at different sites within a municipality were available for an investigation of the effects of chlorapatite vs. fluorapatite treatments on the lead content of groundwater run-off from each site. Three plots were allocated at random to each of these two mineral treatments and the remaining three plots were used as a control. After six months a sample containing twelve replicates of groundwater run-off from each area was analyzed for lead content.

This is called a **nested** or **hierarchical** design. Three plots are nested within each treatment (Figure 14.7).

This design is not appropriate for analysis using a single-factor ANOVA with type of treatment (the apatites) as the factor and the response within each treatment as the number of replicates, because this ignores the presence of the plots that may contribute to the variation within the experiment. You may also be thinking that the design appears pseudoreplicated in that the real level of replication within each treatment is the number of plots rather than the twelve water replicates taken from within each plot. This is true and the nested analysis described below takes this into account.

This design is also unsuitable for analysis as a two-factor ANOVA with mineral treatment as the first factor and plots as the second, because the three plots are simply random subgroups nested within each treatment, which do not intentionally contain different treatment levels of a second factor. For example, the first plot in treatment 1 does not share an exclusive property with the first plot in treatments 2 and 3 (Table 14.3).

Table 14.3 A hierarchical design should not be analyzed as an independent factor design. (a) Correct hierarchical plan for the nested experimental design described in Figure 14.7, and (b) incorrect orthogonal two-factor plan because the plots do not contain different levels of a second factor.

(a) A hierarchical design has one factor nested within the other. The plots have been chosen at random and are nested within each treatment.

No treatment (control)	Chlorapatite	Fluorapatite
Plot C Plot E Plot I	Plot B Plot F Plot G	Plot A Plot D Plot H

(b) Incorrect format of the nested design shown above in (a) as a fully orthogonal design. There is nothing exclusively shared within any of the rows of plots across treatments so it is incorrect to treat the three rows as three different levels of the factor "plot."

	Treatment		
Plot	None (control)	Chlorapatite	Fluorapatite
First within each treatment	12 replicates (Plot C)	12 replicates (Plot B)	12 replicates (Plot A)
Second within each treatment	12 replicates (Plot E)	12 replicates (Plot F)	12 replicates (Plot D)
Third within each treatment	12 replicates (Plot I)	12 replicates (Plot G)	12 replicates (Plot H)

When one factor (e.g. Factor B) is nested within another (e.g. Factor A) it is often written as Factor B(Factor A). For the nested design above, where Factor A is the type of apatite used for the treatment and Factor B is the plots, the following will contribute to the lead content of the groundwater:

$$\text{Lead in groundwater} = \text{Factor A} + \text{Factor B(Factor A)} + \text{error}$$

This is the same as Equation (10.1) for a single-factor ANOVA apart from an additional source of variation from the plots nested within each type of apatite treatment. There is no interaction term because the design is not orthogonal.

A nested ANOVA isolates the effects of treatments and subgroups within these treatments and gives an F ratio for both factors.

14.6 A pictorial explanation of a nested ANOVA

For simplicity the following example has two treatments and two plots nested within each treatment, with only four groundwater replicates measured in

Table 14.4 Data for the lead content (in ppm) of groundwater tested three months after treatment with (a) chlorapatite and (b) fluorapatite. Two plots are nested within each treatment.

Treatment			
chlorapatite		fluorapatite	
Plot 1	Plot 2	Plot 3	Plot 4
30	60	80	110
35	65	85	115
45	75	95	125
50	80	100	130

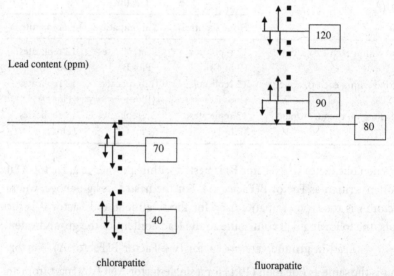

Lead content (ppm)

Figure 14.8 Arrows show the displacement of each replicate from its cell mean, which is the variation due to error only. The number of degrees of freedom is the sum of one less than the number within each of the cells. In this example there are 12 degrees of freedom.

each plot. The data are in Table 14.4. The type of treatment (chlorapatite vs. fluorapatite) is Factor A and the plots are Factor B(A). Figure 14.8 shows the data for each of the four groups in Table 14.4 graphed as four separate cells, including each cell mean and the grand mean.

First, error is estimated. The value for each replicate is displaced from its cell mean by error only (Figure 14.8). The sum of squares for error is obtained by squaring each displacement and adding these together. This quantity is divided by the appropriate degrees of freedom (the sum of one less than the number of replicates within each of the cells) to give the mean square for error.

Second, the subgroups (in this case the plots) are ignored and new means are calculated by combining all of the replicates within each treatment (in this case the type of apatite used) (Figure 14.9). This will give the effect of treatment, but for a nested ANOVA each treatment mean will be displaced from the grand mean because of the **effect of treatment plus the subgroups nested within each treatment, plus error**.

This seems inconsistent with the explanation given for an orthogonal two-factor ANOVA where ignoring a factor (e.g. Factor B) removed it as a source of variation, allowing the effect of the other (e.g. Factor A) to be estimated. For a two-factor orthogonal design, all levels of Factor A are

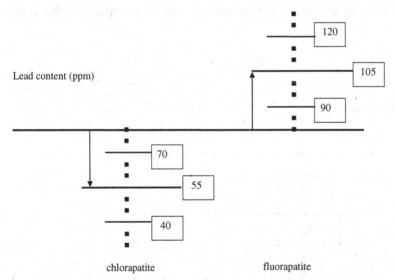

Figure 14.9 Estimation of the effects of Factor A (treatment). The displacement of each combined treatment mean for Factor A from the grand mean shown by the arrows is caused by the average effects of that treatment, plus plots nested within each treatment, plus error. The number of degrees of freedom will be one less than the number of treatments, so in this example with two treatments there is one degree of freedom.

present within every level of Factor B and vice versa, so each of the two factors can be ignored in turn and the effect of each factor separately estimated. For a nested design, however, the effects of Factor B (the subgroups) cannot be excluded in this way because **different** subgroups (here different plots) are present and may contribute very different amounts of variation within each of the levels of Factor A.

The displacements of each treatment mean from the grand mean are squared, multiplied by the number of replicates within their respective treatment and added together to give the sum of squares for Factor A, which will include treatment plus subgroups(treatment) plus error. The number of degrees of freedom is one less than the number of treatments and dividing the sum of squares by this number will give the mean square for Factor A (i.e. treatment plus subgroups(treatment) plus error).

Third, a mean is also calculated for Factor B(A) which is the variation contributed by each subgroup (in this case each plot) (Figure 14.10). Each subgroup mean will only be displaced from its respective treatment mean by the effect of the subgroups plus error. The displacements are squared, multiplied by the number of replicates within their respective subgroups

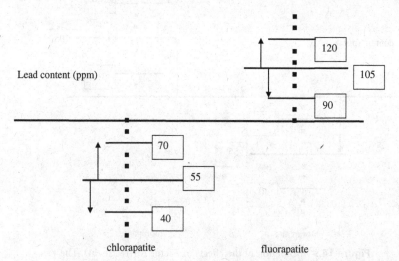

Figure 14.10 Estimation of the effect of Factor B(A). The displacement of each cell mean from its treatment mean is shown by each arrow and is caused by the average effect of that subgroup (each plot) plus error. The number of degrees of freedom will be the sum of one less than the number of plots within each treatment. In this example there are two degrees of freedom.

Table 14.5 The appropriate division and components of each mean square term used to estimate the effect of each factor when Factor B is nested within Factor A.

Source of variation	Calculation of F ratio	Components of each mean square
Factor A (treatment)	$\dfrac{\text{Mean square for Factor A}}{\text{Means square for B(A)}}$	$\dfrac{\text{Factor A+Factor B(A) + error}}{\text{Factor B(A) + error}}$
Factor B(A)	$\dfrac{\text{Mean square for B(A)}}{\text{Mean square error}}$	$\dfrac{\text{Factor B(A) + error}}{\text{Mean square error}}$

and added together to give the Factor B(A) sum of squares. The number of degrees of freedom will be the sum of one less than the number of subgroups within each treatment. Dividing the sum of squares by this number will give the mean square for Factor B(A) (i.e. subgroups + error).

The procedures shown in Figures 14.8 to 14.10 give three separate sums of squares and mean squares:

(a) Factor A: treatment + subgroups(treatment) + error (Figure 14.9)
(b) Factor B(A): subgroups(treatment) + error (Figure 14.10)
(c) error (Figure 14.8)

and no other mean squares are needed to isolate the effects of the treatments from the subgroups nested within each treatment.

First, to isolate the effect of treatment only, the MS for treatment + subgroups(treatment) + error is divided by the MS for subgroups(treatment) + error. Second, to isolate the variation due to subgroups(treatment), the MS for subgroups(treatment) + error is divided by the MS error (Table 14.5).

In the example shown in Figures 14.8 to 14.10, the F ratio for the effect of Factor A will only have one and two degrees of freedom, despite the fact that the experiment used 16 measurements of lead in groundwater. This is appropriate because the level of replication for this comparison is the plots rather than the groundwater replicates taken from within each plot.

Most statistical packages will do a nested ANOVA and the results will be in a similar format to Table 14.6, which gives the results for the data in Table 14.4. If the treatment factor is fixed and significant you are likely to want to carry out a posteriori testing to examine which treatment means are significantly different. The Tukey test (Equation (14.1)) can be used, but

Table 14.6 Results of a nested ANOVA on the data in Table 14.4. Note that the F ratio for the treatment (type of apatite) has been obtained by dividing the MS for type of apatite by the MS for plot.

Source of variation	Sum of squares	df	Mean square	F	P
Apatite type	10000.0	1	10000.0	5.556	0.143
Plot(Apatite)	3600.0	2	1800.0	21.600	0.000
Error	1000.0	12	83.3		

when comparing among treatments the appropriate "MS error" to use in Equation (14.3) is the MS for subgroups (treatments) instead of the error. We suggest you use a more advanced text (e.g. Sokal and Rohlf, 1995 or Zar, 1996) if you need to do a posteriori testing after a nested ANOVA.

This example is the simplest case of a nested or hierarchical design. More complex designs can include several levels of nesting, and nested factors in combination with two- and higher-factor ANOVAs. If you need to use more complex designs it is important to read an advanced text or talk to a statistician before doing the experiment.

14.7 A final comment on ANOVA: this book is only an introduction

Even though this book has five chapters about analysis of variance, it is only an introduction to an enormous and diverse topic. There are far more complex ANOVA models, including those for analyzing repeated measures on the same experimental or sampling unit over time, several variables measured on the same unit, and designs with several factors that include nesting. Hopefully the introduction developed here will make it easier for you to understand more complex designs described in advanced texts!

14.8 Questions

(1) The table below gives the concentration of total polycyclic aromatic hydrocarbons (PAHs) in three benthic sediment cores taken 3 km, 2 km and 1 km from an oil refinery situated on the edge of an estuary. (a) Analyze the data as a two-factor ANOVA without replication, using

depth and distance as factors. Is there a significant effect of distance? Is there a significant effect of depth in the sediment? (b) The marine geoscientist who collected these data mistakenly analyzed them using a single-factor ANOVA comparing the three different cores but ignoring depth (i.e. the table below was simply taken as three columns giving independent data). Repeat this incorrect analysis. Is the result significant? What might be the implications, in terms of the conclusion drawn about the concentrations of PAHs and distance from the refinery, if this were done?

	Distance		
Depth (m)	3 km	2 km	1 km
1	1.11	1.25	1.28
2	0.84	0.94	0.95
3	2.64	2.72	2.84
4	0.34	0.38	0.39
5	4.21	4.20	4.23

(2) A glaciologist who wanted to compare the weight of sediment deposited per square meter in two glacial lakes chose three locations at random within each lake and deployed four sediment traps at each, using a total of 24 traps. This design is summarized below.

Location	Number of traps
First location in lake 1	4 traps
Second location in lake 1	4 traps
Third location in lake 1	4 traps
First location in lake 2	4 traps
Second location in lake 2	4 traps
Third location in lake 2	4 traps

The glaciologist said "I have a two-factor design, where the lakes are one factor and the trap grouping is the second, so I will use a two-factor ANOVA with replication." (a) Is this appropriate? What analysis would you use for this design?

15 | Relationships between variables: linear correlation and linear regression

15.1 Introduction

Often earth scientists obtain data for a sample where two or more variables have been measured on each sampling or experimental unit, because they are interested in whether these variables are **related** and, if so, the **type of functional relationship** between them.

If two variables are related they **vary together** – as the value of one variable increases or decreases, the other also changes in a consistent way.

If two variables are **functionally related,** they vary together and the value of one variable can be predicted from the value of the other.

To detect a relationship between two variables, both are measured on each of several subjects or experimental units and these **bivariate data** examined to see if there is any pattern. One way to do this, by drawing a scatter plot with one variable on the X axis and the other on the Y axis, was described in Chapter 3. Although this can reveal patterns, it does not show whether two variables are **significantly related**, or have a **significant functional relationship**. This is another case where you have to use a statistical test, because an apparent relationship between two variables may only have occurred by chance in a sample from a population where there is no relationship. A statistic will indicate the strength of the relationship, together with the probability of getting that particular result, or an outcome even more extreme, in a sample from a population where there is **no relationship** between the two variables.

Two parametric methods for statistically analyzing relationships between variables are **linear correlation** and **linear regression**, both of which can be used on data measured on a ratio, interval or ordinal scale. Correlation and regression have very different uses, and there have been many cases where correlation has been inappropriately used instead of regression and vice

194

versa. After contrasting correlation and regression, this chapter explains correlation analysis. Regression analysis is explained in Chapter 16.

15.2   Correlation contrasted with regression

Correlation is an **exploratory** technique used to examine whether the values of two variables are significantly **related,** meaning whether the values of both variables change together in a consistent way. (For example, an increase in one may be accompanied by a decrease in the other.) **There is no expectation that the value of one variable can be predicted from the other, or that there is any causal relationship between them.**

In contrast, regression analysis is used to **describe the functional relationship** between two variables so that the value of one can be predicted from the other. A functional relationship means that the value of one variable (called the **dependent** variable, Y) has some relationship to the other (called the **independent** variable, X) in that it is reasonable to hypothesize the value of Y might be affected by an increase or decrease in X, but the reverse is not true. For example, the amount of pitting on limestone buildings is caused by dissolution resulting from acid rain and is likely to be affected by the age of the building because older stones have been exposed to the elements for longer. The opposite is not true – the age of the building is not affected by weathering! Nevertheless, although the amount of weathering is dependent on the age of the building it is **not caused** by age – it is actually caused by acid rain. This is an important point. Regression analysis can be used provided there is a good reason to hypothesize that the value of one variable (the dependent one) is likely to be affected by another (the independent one), but it does not necessarily have to be caused by it.

Regression analysis provides an equation that describes the **functional relationship** between two variables and which can be used to predict values of the dependent variable from the independent one. The very different uses of correlation and regression are summarized in Table 15.1.

15.3   Linear correlation

The Pearson correlation coefficient, symbolized by ρ (the Greek letter rho) for a population and by r for a sample, is a statistic that indicates the extent to

Table 15.1 A contrast between the uses of correlation and regression.

Correlation	Regression
Exploratory – are two variables significantly related?	Definitive – what is the functional relationship between variable Y and variable X and is it significant?
	Predictive – what is the value of Y given a particular value of X?
Neither Y nor X has to be dependent upon the other variable. Neither variable has to be determined by the other.	Variable Y is dependent upon X. It must be plausible that Y is determined by X, but Y does not necessarily have to be caused by X.

which two variables are linearly related, and can be any value from –1 to +1. Usually the population statistic ρ is not known, so it is estimated by the sample statistic r.

An r of +1, which shows a perfect positive linear correlation, will only be obtained when the values of both variables increase together and lie along a straight line (Figure 15.1(a)). Similarly, an r of –1, which shows a perfect negative linear correlation, will only be obtained when the value of one variable decreases as the other increases and the points also lie along a straight line (Figure 15.1(b)). In contrast, an r of zero shows the lack of a relationship between two variables and Figure 15.1(c) gives one example where the points lie along a straight line parallel to the X axis. When the points are more scattered but both variables tend to increase together, the values of r will be between zero and +1 (Figure 15.1(d)), while if one variable tends to decrease as the other increases, the value of r will be between zero and –1 (Figure 15.1(e)). If there is no relationship and considerable scatter (Figure 15.1(f)) the value of r will be close to zero. Finally, it is important to remember that linear correlation will only detect a linear relationship between variables – even though the two variables shown in Figure 15.1(g) are obviously related the value of r will be close to zero.

15.4 Calculation of the Pearson r statistic

A statistic for correlation needs to reliably describe the strength of a linear relationship for any bivariate data set, even when the two variables have

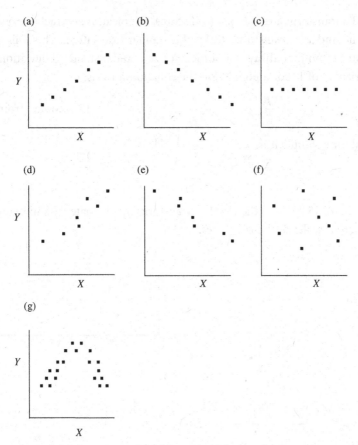

Figure 15.1 Some examples of the value of the correlation coefficient *r*. (a) A perfect linear relationship where $r = 1$, (b) a perfect linear relationship where $r = -1$, (c) no relationship ($r = 0$), (d) a positive linear relationship with $0 < r < 1$, (e) a negative linear relationship where $-1 < r < 0$, (f) no linear relationship (*r* is close to zero) and (g) an obvious relationship but one that will not be detected by linear correlation (*r* will be close to zero).

been measured on very different scales. For example, the values of one variable might range from zero to 10, while the other might range from zero to 1000. To obtain a statistic that always has a value between 1 and −1, with these maximum and minimum values indicating a perfect positive and negative linear relationship respectively, you need a way of standardizing the data. This is straightforward and is done by transforming the values of both variables to their *Z* scores, as described in Chapter 7.

To transform a set of data to Z scores, the mean is subtracted from each value and the result divided by the standard deviation. This will give a distribution that **always** has a mean of zero and a standard deviation (and variance) of 1. For a population the equation for Z is:

$$Z = \frac{X_i - \mu}{\sigma}$$
(15.1 copied from 7.3)

and for a sample it is:

$$Z = \frac{X_i - \overline{X}}{s}$$
(15.2)

Figure 15.2 shows the effect of transforming bivariate data measured on different scales to their Z scores.

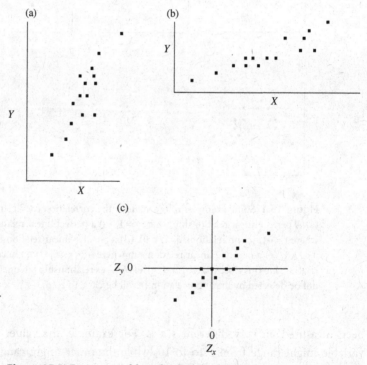

Figure 15.2 For any set of data, dividing the distance between each value and the mean by the standard deviation will give a mean of zero and a standard deviation (and variance) of 1.0. The scales on which X and Y have been measured are very different for cases (a) and (b) above, but transformation of both variables gives the distribution shown in (c) where both Z_x and Z_y have a mean of zero and a standard deviation of 1.0.

Once the data for both variables have been converted to their Z scores, it is easy to calculate a statistic that indicates the strength of the relationship between them.

If the two increase together, large positive values of Z_x will always be associated with large positive values of Z_y and large negative values of Z_x will also be associated with large negative values of Z_y (Figure 15.3(a)).

If there is no relationship between the variables all of the values of Z_y will be zero (Figure 15.3(b)).

Finally, if one variable decreases as the other increases, large positive values of Z_x will be consistently associated with large negative values of Z_y and vice versa (Figure 15.3(c)).

This gives a way of calculating a comparative statistic that indicates the extent to which the two variables are related. If the Z_x and Z_y scores for each of the units are multiplied together and summed (Equation (15.3)), data with a positive correlation will give a total with a positive value, while data with a negative correlation will give a total with a negative one. In contrast, data for two variables that are not related will give a total close to zero:

$$\sum_{i=1}^{n} (Z_{xi} \times Z_{yi}) \tag{15.3}$$

Importantly, the largest possible positive value of $\sum_{i=1}^{n} (Z_{xi} \times Z_{yi})$ will be obtained when each pair of data has exactly the same Z scores for both variables (Figure 15.3(a)) and the largest possible negative value will be obtained when the Z scores for each pair of data are the same number but opposite in sign (Figure 15.3(c)). If the pairs of scores do not vary together completely in either a positive or negative way the total will be a smaller positive (Figure 15.3(d) or negative number (Figure 15.3(f)).

This total will increase as the size of the sample increases, so dividing by the degrees of freedom (N for a population and $n - 1$ for a sample) will give a statistic that has been "averaged," just as the equations for the standard deviation and variance of a sample are averaged and corrected for sample size by dividing by $n - 1$. The statistic given by Equation (15.4) is the Pearson correlation coefficient r.

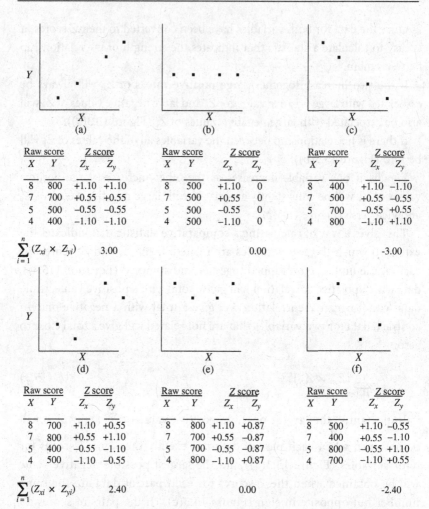

Figure 15.3 Examples of raw scores and Z scores for data with (a) a perfect positive linear relationship (all points lie along a straight line), (b) no relationship, (c) a perfect negative linear relationship (all points lie along a straight line), (d) a positive relationship, (e) no relationship, and (f) a negative relationship. Note that the largest positive and negative values for the sum of the products of the two Z scores for each point occur when there is a perfect positive or negative relationship, and that these values (+3 and –3) are equivalent to $n - 1$ and $- (n - 1)$ respectively.

$$r = \frac{\sum\limits_{i=1}^{n}(Z_{xi} \times Z_{yi})}{n-1} \tag{15.4}$$

More importantly, Equation (15.4) gives a statistic that will only ever be between −1 and +1. This is easy to show. In Chapter 7 it was described how the Z distribution always has a mean of zero and a standard deviation (and variance) of 1.0. If you were to calculate the variance of the Z scores for only one variable you would use the equation:

$$s^2 = \frac{\sum\limits_{i=1}^{n}(Z_i - \overline{Z})^2}{n-1} \tag{15.5}$$

but because $\overline{Z}$ is zero, this equation becomes:

$$s^2 = \frac{\sum\limits_{i=1}^{n}Z_i^2}{n-1} \tag{15.6}$$

and because s^2 is always 1 for the Z distribution, the numerator of Equation (15.6) is always equal to $n-1$.

Therefore, for a set of bivariate data where the two Z scores within each experimental unit are **exactly the same in magnitude and sign**, the equation for the correlation between the two variables:

$$r = \frac{\sum\limits_{i=1}^{n}(Z_{xi} \times Z_{yi})}{n-1} \tag{15.7}$$

will be equivalent to:

$$r = \frac{\sum\limits_{i=1}^{n}Z_{xi}^2}{n-1} \quad \text{or} \quad \frac{n-1}{n-1} = 1.0 \tag{15.8}$$

Consequently, when there is perfect agreement between Z_x and Z_y for each point, the value of r will be 1.0. If the Z scores generally increase together but not all the points lie along a straight line, the value of r will between zero and 1 because the numerator of Equation (15.8) will be less than $n-1$.

Similarly, if every Z score for the first variable is the **exact negative equivalent** of the other, the numerator of Equation (15.8) will be the negative equivalent of $n-1$ so the value of r will be -1.0. If one variable decreases while the other increases but not all the points lie along a straight line, the value of r will be between -1.0 and zero.

Finally, for a set of points along any line parallel to the X axis, all of the Z scores for the Y variable will be zero, so the value of the numerator of Equation (15.6) and r will also be zero.

15.5   Is the value of r statistically significant?

Once you have calculated the value of r, you need to establish whether it is significantly different from zero. Statisticians have calculated the distribution of r for random samples of different sizes taken from a population where there is no correlation between two variables. When $\rho = 0$, the distribution of values of r for many samples taken from that population will be normally distributed with a mean of zero. Both positive and negative values of r will be generated by chance and 5% of these will be greater than a positive critical value or less than its negative equivalent. The critical value will depend on the size of the sample, and as sample size increases the value of r is likely to become closer to the value of ρ. Statistical packages will calculate r and give the probability the sample has been taken from a population where $\rho = 0$.

15.6   Assumptions of linear correlation

Linear correlation analysis assumes that the data are random representatives taken from the larger population of values for each variable, which are normally distributed and have been measured on ratio, interval or ordinal scales. A scatter plot of these variables will have what is called a **bivariate normal distribution**. If the data are not normally distributed, have been measured on a nominal scale only or the relationship does not appear to be linear, they may be able to be analyzed by a non-parametric test for correlation, which is described in Chapter 19.

15.7   Conclusion

Correlation is an exploratory technique used to test whether two variables are related. It is often useful to draw a scatter plot of the data to see if there is any pattern before calculating the correlation coefficient, since the variables may be related together in a non-linear way. The Pearson correlation coefficient is a statistic that shows the extent to which two variables are linearly related, and can have a value between –1.0 and 1.0, with these extremes showing a perfect negative linear relationship and perfect positive linear relationship respectively, while zero shows no relationship. The value of r indicates the way in which the variables are related, but the probability of getting a particular r value is needed to decide whether the correlation is statistically significant.

15.8 Questions

(1) (a) Add appropriate words to the following sentence to specify a **regression analysis**. "I am interested in finding out whether the shell weight of the fossil snail *Littoraria articulata*...................... shell length."
(b) Add appropriate words to the following sentence to specify a **correlation analysis**. "I am interested in finding out whether the shell weight of the fossil snail *Littoraria articulata*........................shell length."

(2) Run a correlation analysis on the following set of 10 bivariate data, given as the values of (X, Y) for each unit: (1,5) (2,6) (3,4) (4,5) (5,5) (6,4) (7,6) (8,5) (9,6) (10,4). (a) What is the value of the correlation coefficient? (You might draw a scatter plot of the data to help visualize the relationship.) (b) Next, modify some of the Y values only to give a highly significant positive correlation between X and Y. Here a scatter plot might help you decide how to do this. (c) Finally, modify some of the Y values only to give a highly significant negative correlation between X and Y.

16 | Linear regression

16.1 Introduction

This chapter explains simple linear regression analysis. The different uses of correlation and regression were contrasted in Chapter 15. Correlation examines if two variables are related. Regression describes the **functional relationship** between a dependent and an independent variable.

16.2 Linear regression

Linear regression analysis is often used by earth scientists. For example, the equation for the regression of one variable on another may suggest hypotheses about why the two variables are related. More practically, regression can be used in situations where the dependent variable is difficult, expensive or impossible to measure, but its values can be predicted from another easily measured variable to which it is functionally related. Here is an example.

It can be quite difficult to measure the temperature of an erupting magma. *In situ* measurements can be made if you can safely get close enough to lower a sheathed thermocouple into the hot lava, but this is a dangerous undertaking. Optical pyrometers can be used to estimate magma temperatures from any distant position with a direct line of sight, but corrections for distance, elevation and air temperatures must be applied.

Fortunately, it has been shown that the SiO_2 content of the magma varies inversely with eruption temperature: basaltic magmas tend to erupt at hotter temperatures (~1200–1300 °C) and more silicic ones at cooler temperatures (~700–800 °C). So eruption temperature can be predicted from the SiO_2 content of the cooled magma, which can be accurately (and safely) measured in the laboratory after the eruption. Temperature is therefore dependent upon (but not caused by) the SiO_2 content and can be predicted

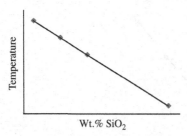

Figure 16.1 An example of the use of regression. The eruption temperature of a magma (the dependent variable) can be predicted by measuring its SiO_2 content (the independent variable). Thus the temperature is determined by and easy to predict from the independent variable, but is not caused by it.

from it by using a regression line (Figure 16.1). This is another deliberate example where the dependent variable is not caused by the independent variable but is plausibly dependent on it.

A linear regression analysis gives an equation for a line that describes the functional relationship between two variables and tests whether the statistics that describe this line are significantly different from zero.

The simplest functional relationship between a dependent and independent variable is a straight line. Only two statistics, the intercept a (which is the value of Y when X is zero) and the slope of the line b, are needed to uniquely describe where that line occurs on a graph.

The position of any point on a straight line can be described by the equation:

$$Y_i = a + bX_i \tag{16.1}$$

where a is the value of Y when $X = 0$, and b is the slope of the line. For example, the equation $Y = 6 + 0.5X$ means "The Y value is 6 units plus half the value of X." Therefore, for this line, when $X_i = 0$, $Y_i = 6$, and when $X_i = 10$, $Y_i = 11$.

Simple linear regression analysis gives an equation for a straight line that is the "best fit" through a set of data points. It is very easy to obtain a and b if all the points lie on a straight line. When the points are scattered, the method for obtaining these statistics is also straightforward.

16.3 Calculation of the slope of the regression line

The slope of the regression line is the amount by which the value of Y increases in relation to an increase in the value of X. For example, if an

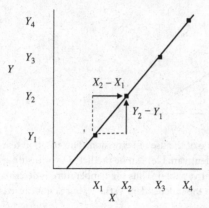

Figure 16.2 Calculation of the slope when all points lie along a straight line. The vertical arrow shows the relative change in Y from Y_1 to Y_2 that occurs with an increase in X from X_1 to X_2 shown by the horizontal arrow. For any two points, $Y_2 - Y_1$ divided by $X_2 - X_1$ will give the slope, which in this case is positive because Y increases as X increases and vice versa.

increase in the value of X by one unit is also accompanied by a one unit increase in the value of Y, the slope of the line is 1.0. If, however, the value of Y decreases by three units for every one unit increase in X then the slope is −3.0.

If all points lie along a straight line, you can calculate the slope by taking any two points and using the equation:

$$b = \frac{Y_2 - Y_1}{X_2 - X_1} \tag{16.2}$$

that divides the relative change in Y by the relative change in X.

Equation (16.2) will not work for a set of points that are scattered. To calculate the slope of the **line of best fit** running through a set of scattered points, a procedure is needed that gives the **average slope**, taking into account the values for all of the points. The equation for calculating b, the slope of the regression line, is:

$$b = \frac{\sum\limits_{i=1}^{n}(X_i - \bar{X})(Y_i - \bar{Y})}{\sum\limits_{i=1}^{n}(X_i - \bar{X})(X_i - \bar{X})} \tag{16.3}$$

This is an extension of Equation (16.2). Instead of calculating the change in X and Y from any two data points, Equation (16.3) calculates an average slope using every point in the data set.

First, the means of X and Y are separately calculated. Next, for each data point, the value of X minus its mean is multiplied by the value of Y minus its mean, and these products are summed. This is the numerator of Equation (16.3), which is then divided by the sum of each value of X minus its mean and squared. It is easy to see how Equation (16.3) will give an appropriate average value for the slope. The first examples are for points that lie on straight lines.

For a line with a slope of +1, as X increases by one unit from its mean, the value of Y will also increase by one unit from its mean (and vice versa if X decreases). The difference between any value of X and its mean will always be the same as the difference between any value of Y and its mean, so the numerator and denominator of Equation (16.3) will be the same thus giving a b value of 1.0 (Figure 16.3(a)).

For a line with a slope of +3, as X increases by one unit from its mean, the value of Y will increase by three units from its mean (and vice versa if X decreases). Therefore, the value of the numerator of Equation (16.3) will always be three times the size of the denominator, no matter how many points are included, thus giving a b value of 3.0 (Figure 16.3(b)).

For a line with a slope of – 1, as X increases by one unit from its mean, the value of Y will decrease by one unit from its mean (and vice versa if X decreases). Therefore the numerator of Equation (16.3) will give a total that is the same magnitude but the negative of the denominator, thus giving a b value of –1.0 (Figure 16.3(c)).

For a line with a slope of – 3, as X increases by one unit from its mean, the value of Y will decrease by three units from its mean (and vice versa if X decreases), so the numerator of Equation (16.3) will always have a negative sign and be three times the value of the denominator, thus giving a b value of –3.0.

Finally, for a line running parallel to the X axis, every value of $Y_i - \bar{Y}$ will be zero, so the total of the numerator of Equation (16.3) will also be zero, thus giving a b value of zero (Figure 16.3(d)).

When the data are scattered, Equation (16.3) will also give the **average** change in Y in relation to the increase in X. Figure 16.4 gives an example. First, cases 16.4 (a), (b) and (c) show three lines, each of which has been drawn through two data points. These lines have slopes of 3.0, 2.0 and 1.0 respectively, and the calculation of each b value is given in the box under the graph. In Figure 16.4 (d) the six data points have been combined. Intuitively,

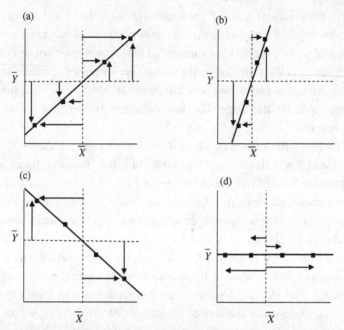

Figure 16.3 Examples of the use of Equation (16.3) to obtain the slope of the regression line. Vertical arrows show $Y_i - \bar{Y}$ and horizontal arrows show $X_i - \bar{X}$. (a) For every point along a line with a slope of 1.0, $Y_i - \bar{Y}$ will be the same magnitude and sign as $X_i - \bar{X}$ so Equation (16.3) will give a value of 1.0. (b) For every point along a line with a slope of 3.0, $Y_i - \bar{Y}$ will be the same sign but three times greater than $X_i - \bar{X}$ so Equation (16.3) will give a value of 3.0. (c) For every point along a line with a slope of –1.0, $Y_i - \bar{Y}$ will be the same magnitude but the opposite sign to $X_i - \bar{X}$ so Equation (16.3) will give a value of –1.0. (d) For a slope of zero, each value of $Y_i - \bar{Y}$ will be zero, so Equation (16.3) will give a value of zero.

this group of six scattered points should have a slope of 2.0, because this is the average of the slopes of the three lines shown in (a), (b) and (c). Equation (16.3) gives this value.

16.4   Calculation of the intercept with the *Y* axis

The intercept of the regression line with the Y axis when $X = 0$ is easy to calculate, using an extension of the formula for the regression line.

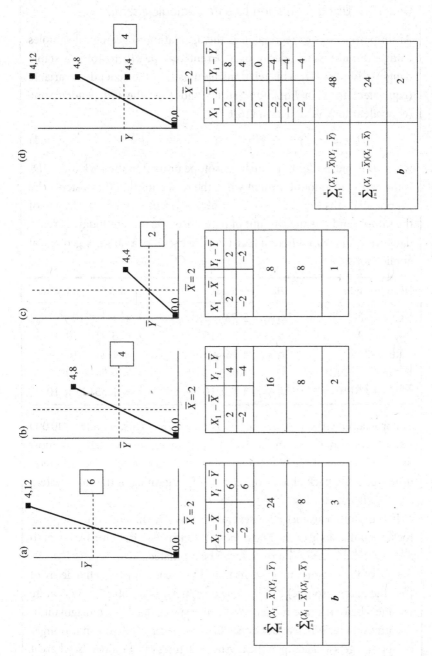

Figure 16.4 Graphs (a), (b) and (c) show three lines of slope 3, 2 and 1 respectively, with two data points on each. The six points have been combined in (d), and the line of best fit through these would be expected to have a slope of 2.0. Use of Equation (16.3) gives this appropriate average for b.

Box 16.1 Using regression to date geological samples

An important use of regression is the age dating of geological samples using the radiometric decay of a parent isotope (P) to form a stable daughter isotope (D). This relationship uses D as the dependent variable (equivalent to Y in Equation (16.1)) and P as the independent one (equivalent to X in Equation (16.1)):

$$D = D_0 + (e^{\lambda t} - 1)P \qquad (16.4)$$

where D is the amount of daughter isotope present in the rock today, D_0 is the expected amount present when the rock cooled in the past, e is the exponential, λ is the decay constant of the parent isotope, t is the age of the sample and P is the amount of parent isotope present today. Parent–daughter pairs with different half-lives are useful in a range of geological applications:

Decay reaction	Half-life (years)	Decay constant (/year)
$^{14}C \rightarrow {}^{14}N$	$t_{1/2} = 5.73 \times 10^3$	$\lambda = 1.2 \times 10^{-4}$
$^{40}K \rightarrow {}^{40}Ar$	$t_{1/2} = 1.3 \times 10^9$	$\lambda = 5.81 \times 10^{-11}$
$^{87}Rb \rightarrow {}^{87}Sr$	$t_{1/2} = 4.86 \times 10^{10}$	$\lambda = 1.42 \times 10^{-11}$
$^{147}Sm \rightarrow {}^{143}Nd$	$t_{1/2} = 1.06 \times 10^{11}$	$\lambda = 6.54 \times 10^{-12}$
$^{238}U \rightarrow {}^{206}Pb$	$t_{1/2} = 4.4 \times 10^9$	$\lambda = 1.55125 \times 10^{-10}$

For example, the half-life of ^{14}C is only 5730 years, so after ~40 000 years there is not enough left to measure, therefore ^{14}C dating is most useful for young rocks and biological remains. Conversely, ancient meteorites that date back billions of years to the beginning of the solar system are usually dated using $^{238}U \rightarrow {}^{206}Pb$.

If D is plotted against P for several crystals of a mineral from the same rock, then the slope of the line of best fit (termed an isochron) is equal to $e^{\lambda t} - 1$, and the age of the rock, t, can be estimated.

One of the most useful systems in isotope geochemistry is the decay of ^{40}K (the parent isotope) to ^{40}Ar (the daughter). The isotope ^{40}Ar is quite volatile when rocks are molten, so it can easily escape from a magma into the atmosphere. When a rock crystallizes, however, the ^{40}Ar can no longer escape and gets locked up in the crystal structures of the minerals, where it accumulates as the ^{40}K in the rock continues to decay. This makes for a very useful geochronometer to tell us the crystallization age, because any subsequent heating back to the liquid state will release all the ^{40}Ar. Thus, if we measure the amounts of ^{40}K and ^{40}Ar in a few feldspar crystals from a rock, we can use the equations above to calculate the crystallization age of the rock and the amount of ^{40}Ar in the original magma (Figure 16.5).

Because

$$Y_i = a + bX_i \qquad\qquad (16.5 \text{ copied from } 16.1)$$

then:

$$\bar{Y} = a + b\bar{X} \qquad\qquad (16.6)$$

and this can be rearranged to give the value of a from:

$$a = \bar{Y} - b\bar{X} \qquad\qquad (16.7)$$

Statistical packages will do this as part of a regression analysis.

16.5 Testing the significance of the slope and the intercept of the regression line

Although the equation for a regression line describes the functional relationship between X and Y, it does not show whether the slope of the line and the intercept are significantly different from zero.

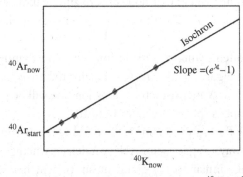

Figure 16.5 A radiometric decay line for ^{40}K $\rightarrow$ ^{40}Ar as measured in four feldspars from the same outcrop. The slope of this line can tell you the age, t, of the rock because λ is known to be 0.581×10^{-10}/yr as measured for this decay reaction. The intercept estimates how much ^{40}K was present in the feldspars when the rock first cooled.

For a population, the equation of the line of best fit is:

$$Y_i = \alpha + \beta X_i \qquad (16.8)$$

but because earth scientists usually only have data for a sample, the population statistics α and β are only estimated by the sample statistics a and b, so you need to test the null hypotheses that a and b are from a population where α and β are zero. Please note that you will find different symbols for the intercept and slope in some texts. Introductory texts generally use a and b (and for a population, α and β) for the intercept and slope, but more advanced texts use b_0 and b_1 for these two sample statistics and β_0 and β_1 for the equivalent population statistics. Here we have used the same symbols as most introductory texts for clarity.

16.5.1 Testing the hypothesis that the slope is significantly different from zero

One method for testing whether the slope of a regression line is significantly different from a slope of zero is very similar to the single-factor ANOVA described in Chapter 10. A pictorial explanation is given in Figures 16.6 and 16.7.

Graphs of four regression lines are shown in Figure 16.6 together with a horizontal line showing $\bar{Y}$, the average value of Y, which the regression line will always cross. If there is no increase or decrease in the value of Y as X increases, the regression line will have a slope of zero and be indistinguishable from the line showing $\bar{Y}$ (Figure 16.6(a)). Nevertheless, samples taken from a population where β is zero will, by chance, have values of b distributed around zero, often giving regression lines that are slightly tilted upwards or downwards (Figure 16.6(b) and (c)). Finally, if there is a marked increase or decrease in Y as X increases, the regression line will be strongly tilted (e.g. a negative slope is shown in Figure 16.6(d)).

The amount by which the regression line is tilted from the horizontal can be detected in the same way a single-factor ANOVA detects whether several treatment means are all similar to the grand mean, or whether any are significantly displaced from it.

In Chapter 10 we described how a single-factor ANOVA calculates an F ratio by dividing the mean square for treatment (i.e. treatment + error) by the mean square for error only. If treatment has no effect, the treatment

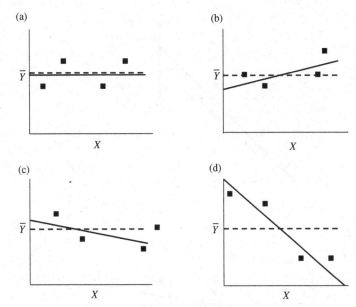

Figure 16.6 A regression line always crosses the line showing $\bar{Y}$. (a) If the slope is exactly zero the regression line will be indistinguishable from the horizontal line showing $\bar{Y}$. Samples from a population where the slope β is zero will nevertheless be expected to include cases with small (b) positive and (c) negative slopes. (d) If Y increases or decreases markedly as X increases, then the regression line will be strongly tilted from the line showing $\bar{Y}$. A negative slope is shown as an example.

means will be the same or close to the grand mean, so the F ratio will be close to 1.0. The test for whether the slope of a regression line is significantly different to the horizontal line showing $\bar{Y}$ is done in a similar way.

First, the regression line will be tilted from the line showing $\bar{Y}$ because of **the variation explained by the regression equation (regression plus error).**

Second, each of the points in the scatter plot will be displaced upwards or downwards from the regression line **because of the remaining variation (error only).**

It is easy to calculate the sums of squares and mean squares for these two separate sources of variation. Figure 16.7 shows scatter plots for two sets of data. The first regression line (16.7(a)) has a large positive slope and the second (16.7(b)) has a slope much closer to zero. The horizontal line on each graph shows $\bar{Y}$. Here you need to think about the vertical displacement

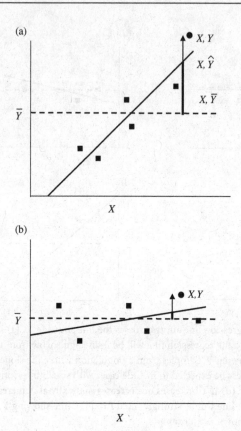

Figure 16.7 (a) The diagonal solid line shows the regression through a scatter plot of six points, and the dashed horizontal line shows $\bar{Y}$. The vertical arrow shows the displacement of one point, symbolized by a circle instead of a square, from $\bar{Y}$. The distance between the point and the Y average $(Y - \bar{Y})$ is the total variation, which can be partitioned into variation explained by the regression line and unexplained variation or error. The heavy part of the vertical line $(\hat{Y} - \bar{Y})$ shows the displacement explained by the regression line (regression plus error) and the remainder $(Y - \hat{Y})$ is unexplained variation (error). Note that (a) when the slope is large, the explained component is also large; and (b) when the slope is close to zero, then the explained component is very small.

of each point from the line showing $\bar{Y}$. To illustrate this, the point at the top far right of each scatter plot in Figure 16.7 has been identified by a circle instead of a square. The vertical arrow running up from $\bar{Y}$ to each of the circled points $(Y - \bar{Y})$ indicates the **total variation** or displacement of that point from $\bar{Y}$. This distance can be partitioned into the two sources of variation mentioned above.

The first is the amount of displacement **explained by the regression line (which is affected by both the regression plus error)** and is the distance $(\hat{Y} - \bar{Y})$ shown by the heavy part of the vertical arrow in Figure 16.7.

The second is the distance $(Y - \hat{Y})$ shown by the lighter vertical part of the arrow in Figure 16.7. This is **unexplained variation** or **error** and often called the **residual variation** because it is the amount of variation remaining between the data points and $\hat{Y}$ that cannot be explained by the regression line.

This gives a way of calculating an F ratio that indicates how much of the variation can be accounted for by the regression.

First, you can calculate the sum of squares for the variation explained by the regression line by squaring the vertical distance between the regression line and $\bar{Y}$ for each point $(\hat{Y} - \bar{Y})$ and adding these together. Dividing this sum of squares by the appropriate number of degrees of freedom will give the mean square due to explained variation (regression plus error).

Second, you can calculate the sum of squares for the unexplained variation by squaring the vertical distance between each point and the regression line $(Y - \hat{Y})$ and adding these together. Dividing this sum of squares by the appropriate number of degrees of freedom will give the mean square due to unexplained variation or "error."

At this stage, you have sums of squares and mean squares for two sources of variation that will be very familiar to you from the explanation of one-factor ANOVA in Chapter 10:

(a) **The variation explained by the regression line (regression plus error).**
(b) **The unexplained residual variation (error only).**

Therefore, to get an F ratio that shows the proportion of the variation explained by the **regression line compared to the unexplained variation due to error, you divide the mean square for (a) by the mean square for (b).**

$$F_{1,n-2} = \frac{\text{MS regression}}{\text{MS residual}} \tag{16.9}$$

If the regression line has a slope close to zero (Figure 16.6(a)) both the numerator and denominator of Equation (16.9) will be similar, so the

---

Box 16.2 A note on the number of degrees of freedom in an ANOVA of the slope of the regression line

The example in Section 16.6 includes an ANOVA table with an F statistic and probability for the significance of the slope of the regression line. Note that the "regression" mean square, which is equivalent to the "treatment" mean square in a single-factor ANOVA, has only one degree of freedom. This is the case for **any** regression analysis, despite the sample size used for the analysis. In contrast, for a single-factor ANOVA the number of degrees of freedom is one less than the number of treatments. This difference needs explaining.

For a single-factor ANOVA, all but one of the treatment means are **free to vary,** but the value of the "final" one is constrained because the grand mean is a set value. Therefore, the number of degrees of freedom for the treatment mean square is always one less than the number of treatments. In contrast, for any regression line every value of $\hat{Y}$ must (by definition) lie on the line. For a regression line of known slope, once the first value of $\hat{Y}$ has been plotted the remainder are no longer **free to vary** because they must lie on the line, so the regression mean square has only one degree of freedom.

The degrees of freedom for error in a single-factor ANOVA are the sum of one less than the number within each of the treatments. Because a degree of freedom is lost for every treatment, if there are a total of n replicates (the sum of the replicates in all treatments) and k treatments, the error degrees of freedom are $n - k$. In contrast, the degrees of freedom for the residual (error) variation in a regression analysis are always $n - 2$. This is because a regression line, which only ever has one degree of freedom, is always only equivalent to an experiment with two treatments.

---

value of the F statistic will be approximately 1.0. As the slope of the line increases (Figure 16.6(b), (c) and (d)), the numerator of Equation (16.9) will become larger, so the value of F will also increase. As F increases, the probability that the data have been taken from a population where the slope of the regression line, β, is zero will decrease and will eventually be less than 0.05. Most statistical packages will calculate the F statistic and give the probability. There is an explanation for the number of degrees of freedom for the F ratio in Box 16.2.

16.5.2 Testing whether the intercept of the regression line is significantly different to zero

The value for the intercept a calculated from a sample is only an estimate of the population statistic α. Consequently, a positive or negative value of a might be obtained in a sample from a population where α is zero. The standard deviation of the points scattered around the regression line can be used to calculate the 95% confidence interval for a, and a single-sample t test can be used to compare the value of a to zero or any other expected value. Once again, most statistical packages include a test to determine if a differs significantly from zero.

16.5.3 The coefficient of determination r^2

The coefficient of determination, symbolized by r^2, is a statistic that shows the **proportion of the total variation of the values of Y from the average $\bar{Y}$ that is explained by the regression line**. It is the regression sum of squares divided by the total sum of squares:

$$r^2 = \frac{\text{Sum of squares explained by the regression}((\text{a}) \text{ above})}{\text{Total sum of squares}((\text{a}) + (\text{b}) \text{ above})} \qquad (16.10)$$

which will only ever be a number from zero to 1.0. If the points all lie along the regression line and it has a slope that is different from zero, the unexplained component (quantity (b)) will be zero and r^2 will be 1. If the explained sum of squares is small in relation to the unexplained, r^2 will be a small number.

16.6   An example: school cancellations and snow

In places at high latitudes, heavy snowfalls are the dream of every young student, because they bring the possibility of school closures (called "snow days"), not to mention sledding, hot chocolate and additional sleep! A school administrator looking for a way to predict the number of school closures on any day in the city of St Paul, Minnesota hypothesized that it would be related to the amount of snow that had fallen during the previous 24 hours. To test this, they examined data from 10 snowfalls. These bivariate data for snowfall (in cm) and the number of school closures on the following day are given in Table 16.1.

Table 16.1 Data for 24-hour snowfall and the number of school closure days for each of 10 snowfalls.

Snowfall (cm)	School closures
3	5
6	13
9	16
12	14
15	18
18	23
21	20
24	32
27	29
30	28

Table 16.2 An example of the table of results from a regression analysis. The value of the intercept a (5.733) is given in the first row, labeled "(Constant)" under the heading "Value". The slope b (0.853) is given in the second row (labeled as the independent variable "Snowfall") under the heading "Value." The final two columns give the results of t tests comparing a and b to zero. These show the intercept, a, is significantly different to zero ($P = 0.035$) and the slope b is also significantly different to zero ($P < 0.001$). The significant value of the intercept suggests that there may be other reasons for school closures (e.g. ice storms, frozen pipes), or perhaps the regression model is not very accurate.

Model	Value	Std error	t	Significance
Constant	5.733	2.265	2.531	0.035
Snowfall	0.853	0.122	7.006	0.001

From a regression analysis of these data a statistical package will give values for the equation for the regression line, plus a test of the hypotheses that the intercept, a, and slope, b are from a population where α and β are zero. The output will be similar in format to Table 16.2.

From the results in Table 16.2 the equation for the regression line is school closures = 5.773 + 0.853 × snowfall. The slope is significantly different to zero (in this case it is positive) and the intercept is also significantly different to zero. You could use the regression equation to predict the number of school closures based on any snowfall between 3 and 30 cm.

Table 16.3 An example of the results of an analysis of the slope of a regression. The significant F ratio shows the slope is significantly different to zero.

	Sum of squares	df	Mean square	F	Significance
Regression	539.648	1	539.648	49.086	0.000
Residual	87.952	8	10.994		
Total	627.600	9			

Most statistical packages will give an ANOVA of the slope. For the data in Table 16.1 there is a significant relationship between school closures and snowfall (Table 16.3).

Finally, the value of r^2 is also given. Sometimes there are two values: r^2, which is the statistic for the sample and a value called "Adjusted" r^2, which is an estimate for the population from which the sample has been taken. The r^2 value is usually the one reported in the results of the regression. For the example above you would get the following values:

$$r = 0.927, \quad r^2 = 0.860, \quad \text{adjusted } r^2 = 0.842$$

This shows that 86% of the variation in school closures with snowfall can be predicted by the regression line.

16.7   Predicting a value of Y from a value of X

Because the regression line has the average slope through a set of scattered points, the predicted value of Y is only the **average** expected for a given value of X. If the r^2 value is 1.0, the value of Y will be predicted without error, because all the data points will lie on the regression line. Usually, however, the points will be scattered around the line. More advanced texts describe how you can calculate the 95% confidence interval for a value of Y and thus predict its likely range.

16.8   Predicting a value of X from a value of Y

Often you might want to estimate a value of the independent variable X from the dependent variable Y. Here is an example. Many elements absorb energy of a very specific wavelength because the movement of electrons or neutrons from one energy level to another within atoms is related to the vibrational modes of crystal lattices. Therefore, the amount of energy absorbed at that wavelength is dependent on the concentration of the element present in a sample. Here it is tempting to designate the concentration of the element as the dependent variable and absorption at the independent one and use regression in order to estimate the concentration of the element present. This is inappropriate because concentration of an element does not depend on the amount of energy absorbed or given off, so one of the assumptions of regression would be violated.

Predicting X from Y can be done by rearranging the regression equation for any point from:

$$Y_i = a + bX_i \tag{16.11}$$

to:

$$X_i = \frac{Y_i - a}{b} \tag{16.12}$$

but here too the 95% confidence interval around the estimated value of X must also be calculated because the measurement of Y is likely to include some error. Methods for doing this are given in more advanced texts.

16.9   The danger of extrapolating beyond the range of data available

Although regression analysis draws a line of best fit through a set of data, it is dangerous to make predictions beyond the measured range of X. Figure 16.8 illustrates that a predicted regression line may not be a correct estimation of the value of Y outside this range.

16.10   Assumptions of linear regression analysis

The procedure for linear regression analysis described in this chapter is often described as a Model I regression, and makes several assumptions.

First, the values of Y are assumed to be from a population of values that are **normally and evenly distributed about the regression line**, with no gross heteroscedasticity. One easy way to check for this is to plot a graph showing the **residuals**. For each data point its vertical displacement on the Y

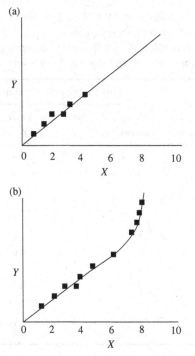

Figure 16.8 It is risky to use a regression line to extrapolate values of Y beyond the measured range of X. The regression line (a) based on the data for values of X ranging from 1 to 5 does not necessarily give an accurate prediction (b) of the values of Y beyond that range. A classic example of such behavior is found in plots of the geothermal gradient of the Earth's interior. At shallow depths, there is generally a linear increase in temperature of ~20 K/km depth, depending on location, but the rate increases as you go deeper into the mantle.

axis either above or below the fitted regression line is the amount of **residual variation** that cannot be explained by the regression line, as described in Section 16.5.1. The residuals are calculated by subtraction (Table 16.4) and plotted on the Y axis, against the values of X for each point and will always give a plot where the regression line is re-expressed as horizontal line with an intercept of zero.

If the original data points are uniformly scattered about the original regression line, the scatter plot of the residuals will be evenly dispersed in a band above and below zero (Figure 16.9). If there is heteroscedasticity the band will vary in width as X increases or decreases. Most statistical packages will give a plot of the residuals for a set of bivariate data.

Table 16.4 Original data and fitted regression line of $Y = 10.8 + 0.9X$. The residual for each point is its vertical displacement from the regression line. Each residual is plotted on the Y axis against the original value of X for that point to give a graph showing the spread of the points about a line of zero slope and intercept.

Original data		Calculated value of $\hat{Y}$ from regression equation	Data for the plot of residuals	
X	Y		Value of X (from original data)	Value of Y $(Y - \hat{Y})$
1	13	11.7	1	1.3
3	12	13.5	3	-1.5
4	14	14.4	4	-0.4
5	17	15.3	5	1.7
6	17	16.2	6	0.8
7	15	17.1	7	-2.1
8	17	18.0	8	-1.0
9	21	18.9	9	2.1
10	20	19.8	10	0.2
11	19	20.7	11	-1.7
12	21	21.6	12	-0.6
14	25	23.4	14	1.6

Second, it is assumed the independent variable X is measured without error. This is often difficult and many texts note that X should be measured with **little** error. For example, levels of an independent variable determined by the experimenter, such as the relative % humidity, are usually measured with very little error indeed. In contrast, variables such as the depth of snowfall from a windy blizzard, or the *in situ* temperature of a violently erupting magma, are likely to be measured with a great deal of error. When the dependent variable is subject to error, a different analysis called Model II regression is appropriate. Again, this is described in more advanced texts.

Third, it is assumed that the dependent variable is determined by the independent variable. This was discussed in Section 16.2.

Fourth, the relationship between X and Y is assumed to be linear and it is important to be confident of this before carrying out the analysis. A scatter plot of the data should be drawn to look for any obvious departures from linearity. In some cases it may be possible to transform the Y variable

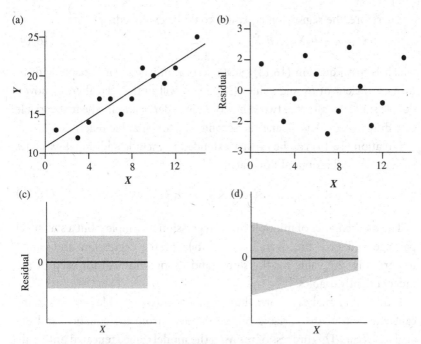

Figure 16.9 (a) Plot of original data in Table 16.4, with fitted regression line $Y = 10.8 + 0.9X$. (b) The plot of the residual $(Y - \hat{Y})$ against the value of X for each data point shows a relatively even scatter about the horizontal line. (c) General form of residual plot for data that are homoscedastic. (d) Residual plot showing one example of heteroscedasticity, where the variance of the residuals decreases with X.

(see Chapter 13) to give a linear relationship and proceed with a regression analysis on the transformed data.

16.11 Multiple linear regression

Multiple linear regression is a straightforward extension of simple linear regression. The simple linear regression equation:

$$Y_i = a + bX_i \qquad \text{(16.13 copied from 16.1)}$$

examines the relationship between the value of a variable Y and another variable X. Often, however, the value of Y might depend upon more than one variable. For example, the sediment yield of a river may be dependent on its drainage area plus other factors such as topographic relief, precipitation and flow rate.

Therefore, the regression equation could be extended:

$$Y_i = a + b_1X_{1i} + b_2X_{2i} \tag{16.14}$$

which is just Equation (16.13) plus a second independent variable with its own coefficient of b_2 and values (X_{2i}). You will notice that there is now a double subscript after the two values of X, in order to specify the first variable (e.g. drainage area) as X_1 and the second (e.g. topographic relief) as X_2.

Equation (16.14) can be further extended to include additional variables such as precipitation and flow rate:

$$Y_i = a + b_1X_{1i} + b_2X_{2i} + b_3X_{3i} + b_4X_{4i} \ldots etc \tag{16.15}$$

The mathematics of multiple linear regression is complex, but a statistical package will give the overall significance of the regression and, more importantly, the value for the slope (and its significance) for each of the independent variables.

If the initial analysis shows that an independent variable has no significant effect on Y, the variable can be removed from the equation and the analysis rerun. This process of **refining the model** can be repeated until only significant independent variables remain, thereby giving the best possible model for predicting Y. There are several procedures for refining, but the one most frequently recommended is to initially include all independent variables, run the analysis and examine the results for the ones that do not appear to affect the value of Y (i.e. variables with non-significant values of b). The least significant is removed and the analysis rerun. This process, called **backward elimination**, is repeated until only significant variables remain.

16.12 Further topics in regression

This chapter is an introduction to linear regression analysis. More advanced analyses include procedures for comparing the slopes and intercepts of two or more regression lines. Non-linear regression models can be fitted to data where the relationship between X and Y is exponential, logarithmic or even more complex. The understanding of simple linear regression developed here is an essential introduction to these methods, which will be discussed further in relation to sequence analysis (Chapter 21).

16.13 Questions

(1) An easy way to help understand regression is to work through a simple contrived example. The set of data below will give a regression with a slope of 0 and an intercept of 10, so the line will have the equation $Y = 10 + 0X$:

X	Y
0	10
0	9
0	11
1	10
1	9
1	11
2	10
2	9
2	11

(a) Use a statistical package to run the regression. What is the value of r^2 for this relationship? Is the slope of the regression significant? (b) Next, modify the data to give an intercept of 20, but with a slope that is still zero. (c) Finally, modify the data to give a negative slope that is significant.

(2) The table below gives data for the weight of alluvial gold recovered from different volumes of stream gravel.

Volume of gravel processed (m^3)	Weight of gold recovered (grams)
1	0.025
2	0.042
3	0.071
4	0.103
5	0.111
6	0.142
7	0.164
8	0.191
9	0.220

(a) Run a regression analysis, where the volume of gravel is the independent variable. What is the value of r^2? Is the relationship significant? What is the equation for the relationship between the weight of gold recovered and the volume of gravel processed? Does the intercept of the regression line differ significantly from zero? Would you expect it to? Why?

17 | Non-parametric statistics

17.1 Introduction

Parametric tests are designed for analyzing data from normally distributed populations. Although these tests are quite robust to departures from normality, and major ones can often be reduced by transformation, there are some cases where the population is so grossly non-normal that parametric testing is unwise. In these cases a powerful analysis can often still be done by using a non-parametric test.

Non-parametric tests are not just alternatives to the parametric procedures for analyzing ratio, interval and ordinal data described in Chapters 8 to 16. Often geoscientists obtain data that have been measured on a nominal scale. For example, Table 3.2 gave data for the locations of 594 tornadoes during the period from 1998–2007 in the southeastern states of the US. This is a sample containing frequencies in several discrete and mutually exclusive categories and there are non-parametric tests for analyzing these types of data (Chapter 18).

17.2 The danger of assuming normality when a population is grossly non-normal

Parametric tests have been specifically designed for analyzing data from populations with distributions shaped like a bell that is symmetrical about the mean with 66.26% of values occurring within $\mu \pm 1$ standard deviation and 95% within $\mu \pm 1.96$ standard deviations (Chapter 7). This distribution is used to determine the range within which 95% of the values of the sample mean, $\overline{X}$, will occur when samples of a particular size are taken from a population. If $\overline{X}$ occurs outside the range of $\mu \pm 1.96$ SEM, the probability the sample has come from that population is less than 5%. If the population is not normally distributed the range occupied by 95% of the values of the mean may be either wider or narrower than assumed, in which case judgments about statistical significance made on the basis of the normal distribution will be misleading.

An example is shown in Figure 17.1. The population is bimodal and the range within which 95% of the values of the means of samples of size $n = 30$

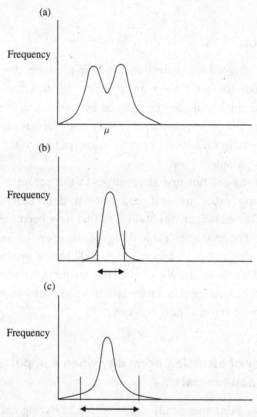

Figure 17.1 Illustration of how the range in which the means of samples from a grossly non-normal population does not correspond to the expected range assuming the population is normally distributed. (a) Distribution of a bimodal population. (b) Actual shape of the distribution of means of sample size $n = 30$ from the population shown in (a). (c) Shape of the distribution of means calculated from the standard error when $n = 30$ assuming the population is normally distributed. Horizontal arrows show the range within which 95% of means would be expected to occur. Note that the expected range in (c) is much wider than the true range in (b).

from this population actually occur is narrower than the range predicted if the population is assumed to be normally distributed.

17.3 The value of making a preliminary inspection of the data

It has already been emphasized that parametric tests for comparing means can often be applied to data from populations that are not normally distributed, because the distribution of the means of samples from most populations will usually be relatively normal (Chapter 6). Once again, however, the example in Section 17.2 emphasizes the value of graphing the data to inspect it for normality and homoscedasticity before attempting a statistical analysis.

The next two chapters describe tests for analyzing nominal scale data, followed by some non-parametric alternatives to the parametric tests for independent and related samples described in Chapters 8 to 12, as well as a non-parametric test for correlation.

18 | Non-parametric tests for nominal scale data

18.1 Introduction

Earth scientists sometimes collect data for which the sampling or experimental units can be assigned to two or more discrete and mutually exclusive categories that are contingent on each other. Consider a paleomagnetic study of the orientation of Earth's magnetic field using cores collected from the sea floor on a traverse perpendicular to the mid-ocean ridge. Roughly 55% of the core samples show the magnetic field pointing north and the remaining 45% show the magnetic field pointing south. These directions are discrete and **mutually exclusive** categories: a rock may be magnetized to point to the north or the south, but it cannot ever be both (the case of an in-between magnetization recorded during a reversal is so rare that it can be considered negligible). These two possibilities, north vs. south, also make up the entire set of possible outcomes and are therefore **contingent** upon each other: for a sample of 100 cores, a decrease in the number in one category (e.g. north-polarized rocks) must be accompanied by an increase in the number in the other (south-polarized rocks) and vice versa.

These are nominal scale data (Chapter 3). The questions researchers ask about these data are the sort asked about any sample(s) from a population.

First, you may want to know the probability that a sample has been taken from a population with a known or expected proportion within each of two or more categories. For example, the field of forensics depends heavily on these types of analyses to build cases based on geological evidence and there are several documented cases where crimes committed on beaches have been solved using knowledge of sand mineralogy (Murray and Tedrow, 1992). Suppose you were hired to examine whether beach sand found in a murder suspect's shoes matched that of the beaches on the south sea island where the victim's body was found. Scientific publications describing that locality give the sand mineralogy at all of the beaches on the island to be 75% coral fragments and 25% basalt grains, and so your null hypothesis is that the sand from the suspect's shoes will also have these proportions. When you examine the sample of 100 grains from the suspect's shoes, you find that it contains 86 coral fragments and 14 basalt grains. Before you go and testify in court, you will need to know the probability that this difference between the observed frequencies in the sample and those expected from the composition of the beach is due to chance.

Second, you may want to know the probability that two or more samples have come from the same population. As an example, consider the handedness of quartz crystals, which is important because of its effect on optical properties. The handedness arises because there are chains of SiO_4 tetrahedra that form a helical spiral around the vertical axis, but the spiral can turn in either a clockwise or counter-clockwise direction. A manufacturer noticed that quartz crystals grown using a new type of alloy in the autoclave tended to be predominantly right-handed. Consequently, 100 quartz crystals grown with the new alloy method and 100 samples grown using the original one were compared. For the new method 67 crystals were right-handed and 33 left-handed, while the original method produced 53 right-handed and 47 left-handed. Here too, the difference between the two samples might be due to chance, or also be affected by the new procedure.

For both of these examples a method is needed that gives the probability of obtaining the observed outcome under the null hypothesis. This chapter describes some tests for analyzing samples of categorical data.

18.2 Comparing observed and expected frequencies: the chi-square test for goodness of fit

The chi-square test for goodness of fit compares the observed frequencies in a sample to those expected in a population. The chi-square statistic is the sum, of each observed frequency minus its expected frequency, squared and then divided by the expected frequency (and was first discussed in Chapter 6):

$$\chi^2 = \sum_{i=1}^{n} \frac{(o_i - e_i)^2}{e_i} \tag{18.1}$$

This is sometimes written as:

$$\chi^2 = \sum_{i=1}^{n} \frac{(f_i - \hat{f}_i)^2}{\hat{f}_i} \tag{18.2}$$

where f_i is the observed frequency and $\hat{f}_i$ is the expected frequency.

It does not matter whether the difference between the observed and expected frequencies is positive or negative because the square of any difference will be positive.

If there is perfect agreement between every observed and expected frequency, the value of chi-square will be zero. Nevertheless, even if the null hypothesis applies, samples are unlikely to always contain the exact proportions present in the population. By chance, small departures are likely and larger departures will also occur, all of which will generate positive values of chi-square. The most extreme 5% of departures from the expected ratio are considered statistically significant and will exceed a critical value of chi-square.

For example, forams can be coiled either counter-clockwise (to the left) or clockwise (to the right). The proportion of forams that coil to the left is close to 0.1 (10%), which can be considered the proportion in the population because it is from a sample of several thousand specimens. A paleontologist, who knew that the proportion of left- and right-coiled forams shows some variation among outcrops, chose 20 forams at random from the same locality and found that four were left-coiled and 16 right-coiled. The question is whether the proportions in the sample were significantly different from the expected proportions of 0.1 and 0.9 respectively. The difference between the population and the sample might be only due to chance, but it might also reflect something about the environment in which the forams lived, such as the water temperature. Table 18.1 gives a worked example of a chi-square test for this sample of left- and right-coiled forams.

The value of chi-square in Table 18.1 has one degree of freedom because the sample size is fixed, so as soon as the frequency of one of the two categories is set the other is no longer free to vary. The 5% critical value of chi-square with one degree of freedom is 3.84 (Appendix A), so the proportions of left- and right-coiled forams in the sample are not significantly different to the expected proportions of 0.1 to 0.9. The chi-square test for goodness of fit can be extended to any number of categories and the degrees of freedom will be $k - 1$ (where k is the number of categories). Statistical packages will calculate the value of chi-square and its probability.

Table 18.1 A worked example using chi-square to compare the observed frequencies in a foram sample to those expected from the known proportions in the population. The observed frequencies in a sample of 20 are 4:16 and the expected frequencies are 2:18.

Coil direction	Left	Right
Observed	4	16
Expected	2	18
Obs – Exp	2	–2
(Obs – Exp)^2	4	4
$\dfrac{(Obs - Exp)^2}{Exp}$	2	0.22

$$\chi^2 = \sum_{I=1}^{n} \frac{(o_i - e_i)^2}{e_i} = 2.22$$

18.2.1 Small sample sizes

When expected frequencies are small, the calculated chi-square statistic is inaccurate and tends to be too large, therefore indicating a lower than appropriate probability which increases the risk of Type 1 error. It used to be recommended that no expected frequency in a chi-square goodness of fit test should be less than five, but this has been relaxed somewhat in the light of more recent research, and it is now recommended that no more than 20% of expected frequencies should be less than five.

An entirely different method, which is not subject to bias when sample size is small, can be used to analyze these data. It is an example of a group of procedures called **randomization tests** that will be discussed further in Chapter 19. Instead of calculating a statistic that is used to estimate the probability of an outcome, a randomization test uses a computer program **to simulate the repeated random sampling of a hypothetical population containing the expected proportions in each category**. These samples will often contain the same proportions as the population, but departures will occur by chance. The simulated sampling is **iterated,** meaning it is repeated, several thousand times and the resultant distribution of the statistic used to identify the most extreme 5% of departures from the expected proportions. Finally, the actual proportions in the real sample are compared to this distribution. If the sample statistic falls within the region where the most

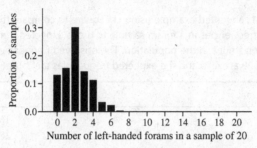

Figure 18.1 An example of the distribution of outcomes from a Monte Carlo simulation where 10 000 samples of size 20 are taken at random from a population containing 0.1 left-coiled and 0.9 right-coiled forams. Note that the probability of obtaining four or more left-coiled forams in a sample of 20 is greater than 0.05.

extreme 5% of departures from the expected occur, the sample is considered significantly different from the population.

Repeated random sampling of a hypothetical population is an example of a more general procedure called the **Monte Carlo method** that uses the properties of the sample, or the expected properties of a population, and takes a large number of simulated random samples to create a distribution that would apply under the null hypothesis.

For the data in Table 18.1, where the sample size is 20 and the expected proportions are 0.1 left-coiled to 0.9 right-coiled, a randomization test works by taking several thousand random samples, each of size 20, from a hypothetical population containing these proportions. This will generate a distribution of outcomes similar to the one shown in Figure 18.1, which is for 10 000 samples. If the procedure is repeated another 10 000 times, then the outcome is unlikely to be exactly the same, but nevertheless will be very similar to Figure 18.1 because so many samples have been taken. It is clear from Figure 18.1 that the likelihood of a sample containing four or more forams with tails coiling to the left is greater than 0.05.

18.3   Comparing proportions among two or more independent samples

Earth scientists often need to compare the proportions in categories among two or more samples to test the null hypothesis that these have come from the same population. Unlike the previous example, there are no expected

Table 18.2 Data for 20 water samples taken at each of three locations to characterize the presence or absence of nitrate contamination.

	Townsville	Bowen	Mackay
Contaminated	12	7	14
Uncontaminated	8	13	6

proportions – instead these tests examine whether the proportions in each category are heterogeneous among samples.

18.3.1 The chi-square test for heterogeneity

Here is an example for three samples, each containing two mutually exclusive categories. Hydrologists managing water aquifers are often concerned about contamination from agricultural fertilizers containing nitrate (NO_3^-), which is a very soluble form of nitrogen that can be absorbed by plant roots. Unfortunately nitrate can leach into groundwater and make it unsafe for drinking. A hydrologist hired to evaluate aquifers in three adjacent rural areas sampled 20 wells in each for the presence/absence of detectable levels of nitrate. The researcher did not have a preconceived hypothesis about the expected proportions of contaminated and uncontaminated aquifers – they simply wanted to compare the three locations. The data are shown in Table 18.2. This format is often called a **contingency table**.

These data are used to calculate an **expected frequency** for each of the six cells. This is done by first calculating the row and column totals (Table 18.3(a)) which are often called the **marginal totals**. The proportions of contaminated and uncontaminated aquifers in the marginal totals shown in the right-hand column of Figure 18.3 are the overall proportions within the sample. Therefore, under the null hypothesis of no difference in nitrate among locations, each will have the same proportion of contaminated wells. To obtain the expected frequency for any cell under the null hypothesis, the column total and the row total corresponding to that cell are multiplied together and divided by the grand total. For example, in Table 18.3(b) the expected frequency of contaminated wells in a sample of 20 from Townsville is $(20 \times 33) \div 60 = 11$ and the expected frequency of uncontaminated wells from Mackay is $(20 \times 27) \div 60 = 9$.

Table 18.3 (a) The marginal totals for the data in Table 18.2. To obtain the expected frequency for any cell, its row and column total are multiplied together and divided by the grand total. (b) Note that the expected frequencies at each location (11:9) are the same and also correspond to the proportions of the marginal totals (33:27).

(a) Observed frequencies and marginal totals.

	Townsville	Bowen	Mackay	Row totals
Contaminated	12	7	14	33
Uncontaminated	8	13	6	27
Column totals	20	20	20	Grand total = 60

(b) Expected frequencies calculated from the marginal totals.

	Townsville	Bowen	Mackay	Row totals
Contaminated	11	11	11	33
Uncontaminated	9	9	9	27
Column totals	20	20	20	Grand total = 60

After the expected frequencies have been calculated for all cells, Equation (18.1) is used to calculate the chi-square statistic. The number of degrees of freedom for this analysis is one less than the number of columns, multiplied by one less than the number of rows, because all but one of the values within each column and each row are free to vary, but the final one is not because of the fixed marginal total. Here, therefore, the number of degrees of freedom is $2 \times 1 = 2$. The smallest contingency table possible has two rows and two columns (this is called a 2×2 table), which will give a chi-square statistic with only one degree of freedom.

18.3.2 The G test or log-likelihood ratio

The **G test** or **log-likelihood ratio** is another way of estimating the chi-square statistic. The formula for the G statistic is:

$$G = 2 \sum_{i=1}^{n} f_i \ln\left(\frac{f_i}{\hat{f}_i}\right) \tag{18.3}$$

This means, "The G statistic is twice the sum of the frequency of each cell multiplied by the natural logarithm of each observed frequency divided by the expected frequency." The formula will give a statistic of zero when each expected frequency is equal to its observed frequency, but any discrepancy will give a positive value of G. Some statisticians recommend the G test and others recommend the chi-square test. There is a summary of tests recommended for categorical data near the end of this chapter.

18.3.3   Randomization tests for contingency tables

A randomization test procedure similar to the one discussed in Section 18.2.1 for goodness-of-fit tests can be used for any contingency table. First, the marginal totals of the table are calculated and give the expected proportions when there is no difference among samples. Then, the Monte Carlo method is used to repeatedly "sample" a hypothetical population containing these proportions, with the constraint that both the column and row totals are fixed. Randomization tests are available in some statistical packages.

18.4   Bias when there is one degree of freedom

When there is only one degree of freedom and the total sample size is less than 200, the calculated value of chi-square has been shown to be inaccurate because it is too large. Consequently it gives a probability that is smaller than appropriate, thus increasing the risk of Type 1 error. This bias increases as sample size decreases, so the following formula, called **Yates' correction** or **the continuity correction**, was designed to improve the accuracy of the chi-square statistic for small samples with one degree of freedom.

Yates' correction removes 0.5 from the **absolute** difference between each observed and expected frequency. (The absolute difference is used because it converts all differences to positive numbers, which will be reduced by subtracting 0.5. Otherwise, any negative values of $o_i - e_i$ would have to be increased by 0.5 to make their absolute size and the square of that smaller.) The absolute value is the positive of any number and is indicated by enclosing the number or its symbol by two vertical bars (e.g. $|-6| = 6$). The subscript "$_{adj}$" after the value of chi-square means it has been adjusted by Yates' correction.

$$\chi^2_{adj} = \sum_{i=1}^{n} \frac{(|o_i - e_i| - 0.5)^2}{e_i} \tag{18.4}$$

From Equation (18.4) it is clear that the compensatory effect of Yates' correction will become less and less as sample size increases. Some authors (e.g. Zar, 1996) recommend that Yates' correction is applied to all chi-square tests having only one degree of freedom, but others suggest it is unnecessary for large samples and recommend the use of the Fisher Exact Test (see Section 18.4.1 below) for smaller ones.

18.4.1 The Fisher Exact Test for 2 × 2 tables

The Fisher Exact Test accurately calculates the probability that two samples, each containing two categories, are from the same population. **This test is not subject to bias** and is recommended when sample sizes are small or more than 20% of expected frequencies are less than five, but it can be used for any 2 × 2 contingency table.

The Fisher Exact Test is unusual in that it does not calculate a statistic that is used to estimate the probability of a departure from the null hypothesis. Instead, the probability is calculated directly.

The easiest way to explain the Fisher Exact Test is with an example. Table 18.4 gives data for the presence or absence of mollusc species with anti-predator adaptations on either side of the Cretaceous/Tertiary (K/T) extinction boundary. A typical adaptation might include development of a thicker, stronger shell, or perhaps a decrease in the size of the aperture (opening) of the shell to discourage shell-peeling by predatory crabs (e.g. Vermeij, 1978). However, during an environmental event causing mass extinction, such adaptations might require more food or reduce mobility, either of which may diminish the species' ability to survive. To test this hypothesis, a paleontologist examined ten outcrops, five below the K/T boundary and five above it. The results for the presence or lack of detection of thick-shelled molluscs are in Table 18.4. These frequencies are too small for accurate analysis using a chi-square test.

Table 18.4 Data for the presence/absence of mollusc species with thick shells in ten samples above and below the mass extinction boundary between the Cretaceous and Tertiary periods. The sample deliberately included five samples above the boundary layer and five below it. The marginal totals show that four samples contain species with thick shells and six do not.

	Above K/T boundary	Below K/T boundary	
Thick-shelled molluscs present	0	4	4
Thick-shelled molluscs not found	5	1	6
Totals	5	5	10

Table 18.5 Under the null hypothesis that there is no effect of mass extinction on the presence of molluscs with thick shells, the expected proportions of rocks with and without thick-shelled molluscs in each sample (2:3 and 2:3) will correspond to the marginal totals for the two rows (4:6). The proportions of samples from above and below the K/T boundary (2:2) and (3:3) will also correspond to the marginal totals for the two columns (5:5).

	Above K/T boundary	Below K/T boundary	
Thick-shelled molluscs present	2	2	4
Thick-shelled molluscs not found	3	3	6
Totals	5	5	10

If there were no effect of mass extinction, then you would expect, under the null hypothesis, that the proportion of samples containing molluscs with thicker shells (representing anti-predatory adaptations) in each locality (above and below the K/T boundary) would be the same as the marginal totals (Table 18.5) with any departures being due to chance. The Fisher Exact Test uses the following procedure to calculate the probability of an outcome equal to or more extreme than the one observed, which can be used to decide whether it is statistically significant.

First, the four marginal totals are calculated, as shown in Table 18.5.

Second, all of the possible ways in which the data can be arranged within the four cells of the 2×2 table are listed, subject to the constraint that the marginal totals must remain unchanged. This is **the total set of possible outcomes for the sample.** For these marginal totals, the most likely outcome under the null hypothesis of no difference between the samples is shown in Table 18.5 and identified as (c) in Table 18.6.

Table 18.6 The total set of possible outcomes for the number of outcrops with and without thick-shelled molluscs, subject to the constraint that there are five outcrops on each side of K/T mass extinction and four have thick-shelled molluscs while six lack them. The most likely outcome, where the proportions are the same both above and below the K/T boundary, is shown in the central box (c). The actual outcome is case (e).

	Above K/T boundary	Below K/T boundary	Above K/T boundary	Below K/T boundary	Above K/T boundary	Below K/T boundary	Above K/T boundary	Below K/T boundary	Above K/T boundary	Below K/T boundary
Thick-shelled molluscs present	4	0	3	1	2	2	1	3	0	4
Thick-shelled molluscs not found	1	5	2	4	3	3	4	2	5	1
	(a)		(b)		(c) Expected under the null hypothesis		(d)		(e) Observed outcome	

For a sample of ten outcrops, five of which are above the K/T boundary and five below, together with the constraint that four outcrops must have thick-shelled molluscs and six must lack them, there are five possible outcomes (Table 18.6). To obtain these, you start with the outcome expected under the null hypothesis (c), choose one of the four cells (it does not matter which) and add one to that cell. Next, adjust the values in the other three cells so the marginal totals do not change. Continue with this procedure until the number within the cell you have chosen cannot be increased any further without affecting the marginal totals. Then go back to the expected outcome and repeat the procedure by subtracting one from the same cell until the number in it cannot decrease any further without affecting the marginal totals (Table 18.6).

Third, the actual outcome is identified within the total set of possible outcomes. For this example, it is case (e) in Table 18.6. The probability of this outcome, together with any more extreme departures in the same direction from the one expected under the null hypothesis (here there are none more extreme than (e)) can be calculated from the probability of getting this particular arrangement within the four cells by sampling a set of ten outcrops, four of which contain thick-shelled molluscs and six of which do not, with the outcrops sampled from above and below the K/T boundary. This is similar to the example used to introduce hypothesis testing in Chapter 6, where you had to imagine a sample of hornblende vs. quartz grains in a beach sand. Here, however, a very small group is sampled without replacement, so the initial probability of selecting an outcrop with thick-shelled molluscs present is 4/10, but if one is drawn, the probability of next drawing an outcrop with thick-shelled molluscs is now 3/9 (and 6/9 without). We deliberately have not given this calculation because it is long and tedious, and most statistical packages do it as part of the Fisher Exact Test.

The calculation gives the exact probability of getting the observed outcome or a more extreme departure in the same direction from that expected under the null hypothesis. This is a one-tailed probability, because the outcomes in the opposite direction (e.g. on the left of (c) in Table 18.6) have been ignored. For a two-tailed hypothesis you need to double the probability. When the probability is less than 0.05, the outcome is considered statistically significant.

Table 18.7 Data for the number of dinosaur footprints found within a 1 m² area on five rock slabs, each of a different rock type. The numbers counted on each slab are not mutually exclusive or contingent upon the numbers within any other, so the data are unsuitable for analysis by a goodness-of-fit test.

	shale	mudstone	siltstone	graywacke	sandstone
Number of footprints	14	1	16	17	2

18.5   Three-dimensional contingency tables

The contingency tables described in this chapter are two-dimensional, but three-dimensional tables can also be analyzed. For example, if you had two or more samples within which two categorical variables have been measured on each unit (e.g. presence/absence of nitrate and aluminum above and below the legal limit), these would give a contingency table consisting of a three-dimensional block of cells with one column and two rows. Three-dimensional chi-square analyses are described in more advanced texts.

18.6   Inappropriate use of tests for goodness of fit and heterogeneity

Tests for goodness of fit and contingency tables assume that the data are mutually exclusive and contingent upon one another. It is also assumed that the categories are the entire set possible within each sample. Occasionally, however, these tests are misused. The most common misuse occurs when samples are incorrectly considered as categories, as shown in the following example.

A group of paleontologists interested in the study of fossilized tracks and traces (paleoichnology) examined fossil footprint collections from 200 million year old rock slabs. To evaluate possible dinosaur migration habits, they selected one square meter on five slabs with different rock types, and counted the number of footprints in each one (Table 18.7). Overall, 50 footprints were found.

The data were analyzed using a chi-square test for goodness of fit, with the null hypothesis that equal numbers of footprints (in this case 10, because

50 were found in total and there were 5 slabs) would be expected on each slab. **Unfortunately, these data are not suitable for a goodness-of-fit test because the five conditions are neither mutually exclusive nor contingent categories within a sample.** This is clear if you consider that any dinosaur could have walked on multiple types of sediments, or that it could have chosen to walk on only one. The numbers in each slab are actually single replicates of ratio scale data.

To avoid the pitfall of confusing categories and samples, you need to ask yourself "Do I have data for categories that are mutually exclusive and contingent within each sample, or are my 'categories' really separate independent samples?"

18.7   Recommended tests for categorical data

Several tests have been developed for data that are frequencies in mutually exclusive and contingent categories. The following are broad recommendations.

When comparing the frequencies in two or more categories within a single sample to their expected proportions, Yates' corrected chi-square can be used where no more than 20% of expected frequencies are less than five. A randomization test can be used for any sized sample.

For 2×2 contingency tables the Fisher Exact Test will give an unbiased probability and is available in most statistical packages.

For contingency tables with more than two rows and columns, the chi-square G test or can be used if no more than 20% of expected frequencies are less than five. A randomization test will give an unbiased probability for any sized sample.

18.8   Comparing proportions among two or more related samples of nominal scale data

If you have measured the same variable more than once on each sampling or experimental unit, then the samples are not independent and need to be analyzed using a test for **related samples**. Table 18.8 gives data for the accretion and growth of 12 beaches on a glacial spit off the coast of Massachusetts. Surveys and photographs were used to document the size of beaches in two consecutive years. In the first year, 11 of the 12 beaches

Table 18.8 The status of 12 ocean beaches, as measured in August of two successive years, preceding and following construction of jetties at the mouth of a nearby river. These two samples are not independent because they contain the same 12 beaches.

Beach	Year 1	Year 2
1	Accreting	Accreting
2	Accreting	Accreting
3	Accreting	Eroding
4	Accreting	Accreting
5	Accreting	Eroding
6	Eroding	Eroding
7	Accreting	Accreting
8	Accreting	Eroding
9	Accreting	Accreting
10	Accreting	Eroding
11	Accreting	Eroding
12	Accreting	Eroding

were accreting sand, adding hundreds of yards of sand to the coastline each year. Over the intervening winter several new jetties were constructed to control and funnel the flow of water and sand, in an attempt to maintain the course of the nearby river channel and improve its navigability for shipping. By the following spring, deposition patterns had dramatically changed: only five beaches continued to accrete sand and the remaining seven were eroding. The null hypothesis was that the proportion of accreting/eroding beaches would be unaffected by the jetties, while the alternate hypothesis was that the proportion of accreting/eroding beaches would be affected. These are two related samples so it is not appropriate to analyze them with a test that compares two independent ones.

The **McNemar test for the significance of changes** compares two related samples of nominal scale data in two categories. The data in Table 18.8 are summarized in a 2 × 2 table giving the number of individuals in all four possible combinations of categories and samples. These are (a) accreting before and after the jetty construction, (b) accreting before and eroding after, (c) eroding before and accreting after and (d) eroding before and eroding after (Table 18.9).

Table 18.9 The status of 12 beaches tested before and after construction of jetties at the mouth of a nearby river. Two cells show the beaches that changed; these are (b) from accreting to eroding and (c) from eroding to accreting. Note that in this example there are no beaches in the second category.

	After	
Before	Accreting	Eroding
Accreting	(a) 5	(b) 6
Eroding	(c) 0	(d) 1

The null hypothesis predicts that there will be no difference in the proportions of eroding beaches between the two samples, while the alternate predicts there will be a difference. Therefore, under the null hypothesis, the beaches that **did** change status (combinations (b) and (c)) would be expected to include equal numbers that changed from eroding to accreting and from accreting to eroding, so you would expect cells (b) and (c) of Table 18.9 to contain equal frequencies. If, however, the proportion of eroding beaches differed before and after construction of the jetties, then the frequencies in these two cells would be expected to be unequal.

In this example, six beaches changed status, so three would be expected to change from eroding to accreting and vice versa. The McNemar test ignores categories (a) and (d) where no change has occurred and compares the observed and expected frequencies in cells (b) and (c) using a goodness-of-fit test (e.g. the chi-square, exact or randomization tests for two mutually exclusive categories discussed earlier in this chapter, or the exact probability calculated from the binomial distribution discussed in Chapter 6). If there is a statistically significant difference between the proportions in these two categories it indicates a change between the two samples.

For three or more related samples of nominal scale data in two categories, the **Cochran Q test** is an extension of the McNemar test. These tests are also included in most statistical packages.

18.9 Questions

(1) The ratio of left-handed to right-handed people in the human population is about 1 : 9. When one of us (SM) was in first year of

university, he was in a tutorial group of 14 that contained 13 left-handers and one right-hander. Students had been assigned to tutorial groups at random. Is this departure from the expected proportion of left-handers significant? What could you conclude about this occurrence?

(2) To help understand how the chi-square statistic is related to departures from expected proportions, it is useful to use a statistical package to compare a sample of contrived data to the expected proportions for a population. Use expected proportions of 1:1 (e.g. left- and right-hand forams) and a sample size of 100. (a) Initially set the sample numbers at 50:50 left to right and run a single-sample chi-square test. What would you expect the value of chi-square to be? (b) Now, change the sample proportions to 45 and 55 and rerun the test. Repeat this for increasing departures from the expected ratio (e.g. 40 and 60, 35 and 65, 30 and 70). What happens to the value of chi-square? What happens to the probability?

(3) A team of paleontologists searched two sandstone outcrops for three hours each and found 8 trilobites in one outcrop and 22 in the other. They compared the numbers using a chi-square test and an expected ratio of 15:15. Was this test appropriate? Please discuss.

(4) Do you think the sampling described in Section 18.8 was well designed? Might there be other possible reasons for the change in status of the 12 beaches from Year 1 to Year 2? How might you improve the experimental design to take other possible influences into account?

19 | Non-parametric tests for ratio, interval or ordinal scale data

19.1 Introduction

This chapter describes some non-parametric tests for ratio, interval and ordinal scale univariate data. These tests do not use the predictable distribution of sample means, which is the basis of most parametric tests, to infer whether samples are from the same population. Consequently non-parametric tests are generally not as powerful as their parametric equivalents, but if the data are **grossly** non-normal and cannot be satisfactorily improved by transformation, it is necessary to use one of these tests.

Non-parametric tests are often called "distribution free tests" but most nevertheless assume that the samples being analyzed are from populations with the same distribution. **Therefore, most non-parametric tests should not be used where there are gross differences in distribution (including the variance) among samples.** The general rule that the ratio of the largest to smallest sample variance should not exceed 4 : 1 discussed in Chapter 13 also applies to non-parametric tests.

Many non-parametric tests for ratio, interval or ordinal data calculate a statistic from a comparison of two or more samples and work in the following way.

First, the raw data are converted to **ranks**. For example, the lowest value is assigned the rank of "1", the next highest "2" etc. This transforms the data to an ordinal scale (see Chapter 3) with the ranks indicating only their relative order. Under the null hypothesis that the samples are from the same population you would expect a similar range of ranks within each, with differences among samples occurring only by chance.

Second, a statistic that reflects any differences in the ranks among samples is calculated and its value compared to the expected distribution of this statistic when samples have been taken from the same population. If the calculated value falls within the range generated by the most extreme 5%

of departures from the null hypothesis the result is considered statistically significant. Most statistical packages give the value of the test statistic, together with the probability of that outcome. Randomization and exact tests can also be used to compare two or more samples of ratio, interval or ordinal scale data and are described in this chapter.

19.2   A non-parametric comparison between one sample and an expected distribution

The Kolmogorov–Smirnov one-sample test can be used to compare the distribution of a single sample of continuous data to an expected or known distribution. Here is an example. Limestone is relatively soft, easily worked, and has an attractive appearance, so it is often cut into blocks and used as building stone and tombstones. Unfortunately, it is composed mostly of calcium carbonate (calcite, or $CaCO_3$) and therefore vulnerable to dissolution by rainwater, which is naturally slightly acidic but can be extremely so (hence the term "acid rain") in polluted areas.

Tombstones in a graveyard east of Detroit, Michigan have a bimodal distribution of the amount of weathering: some stones start to show signs of partial dissolution after only 10 years, while others take much longer. All of these tombstones come from a nearby quarry that has been worked for more than 200 years. The amount of weathering in the tombstones can be quantified by examining the lead that was initially inserted into the carved lettering and then polished flush with the surface of the stone: acid rain will slowly dissolve the limestone but not the lead. Thus, a "lead lettering index" (LLI) that is the absolute difference in height between the lead letters and the eroded surface of the stone around them was created to quantify the weathering of a tombstone.

Generations of geology students have visited the town's graveyards and amassed LLI data along with age of the limestone tombstones. These data have been obtained from such a large number of stones over several decades that they can be assumed to be the distribution for the population.

The owner of the local quarry where all the tombstones originated noticed that some areas of the quarry walls, containing large quantities of fossils, showed very little weathering. The owner wondered if stone from these areas might produce more lasting tombstones, so they commissioned a geologist to test the hypothesis that tombstones containing the most fossils were more resistant to dissolution. The geologist visited a local graveyard and measured

Table 19.1 A worked example of a Kolmogorov–Smirnov one-sample test. The LLIs of 36 fossil-rich gravestones that are 100 years old are summarized as frequencies that are converted to proportions of the sample. These are expressed as cumulative proportions. First each expected cumulative proportion (from the known or expected distribution of the population) is subtracted from the observed cumulative proportion and expressed as the absolute difference D. The greatest value of the statistic is identified. In this case, it is 0.1217 (shown in bold in the far right-hand column). If the probability of obtaining this or a more extreme value of the D statistic is less than 5%, then the two distributions are considered significantly different.

LLI (lead lettering index) in mm	Observed numbers in each category	Observed relative frequency in each category	Observed cumulative proportions in each category	Expected cumulative proportions (from the population)	Absolute difference D (observed F_i minus expected $\hat{F}$)
		f_i	F_i	$\hat{F}_i$	$D = F_i - \hat{F}_i$
<2.0	5	0.1389	0.1389	0.1045	0.0344
2.0–2.49	10	0.2778	0.4167	0.3943	0.0224
2.5–2.99	4	0.1111	0.5278	0.5623	0.0345
3.0–3.49	2	0.0556	0.5833	0.6236	0.0403
3.5–3.99	8	0.2222	0.8056	0.6839	**0.1217**
4.0–4.49	5	0.1389	0.9444	0.9325	0.0119
≥4.50	2	0.0556	1.0000	1.0000	0.0000
Total	36	1.0000			

the LLIs of 36 tombstones, each of which was 100 years old and contained prominent fossils. This distribution was compared to the known distribution of LLIs for the (much larger) population of 100 year old tombstones measured by generations of students. Both sets of data are summarized in Table 19.1.

If you make a preliminary inspection of these data by drawing a histogram you will find the LLI distribution is bimodal and clearly not appropriate for analysis using a parametric test such as a single-sample t test. A non-parametric test is needed.

For a Kolmogorov–Smirnov one-sample test you need to subdivide the range (here the LLI for the tombstones) into several intervals of equal width, and count the number of cases within each. This is the same procedure used for drawing a frequency histogram described in Section 3.3.2 and you should follow those guidelines.

A worked example is given in Table 19.1. First, the number of cases in each interval is converted to relative frequencies (and thus their proportions of the sample). Second, the relative frequencies are progressively added together to give the cumulative relative frequency (i.e. the cumulative proportion). Here too, this procedure is the same as drawing a cumulative frequency graph (Section 3.3.3).

The cumulative proportions for the sample have to be compared to the cumulative proportions of the known distribution for the population. To do this you calculate the **absolute** value of the difference between the cumulative observed relative frequency (F_i) and cumulative expected relative frequency ($\hat{F}_i$) for each interval, which will always be positive:

$$D = |F_i - \hat{F}_i| \tag{19.1}$$

If the observed and expected proportions in each interval are the same the value of D will always be zero. As the discrepancy between the observed and expected proportions increases, the value of D will increase.

The largest value is found and called the D statistic. For the worked example in Table 19.1 the D statistic is 0.1217. Statistical packages generate the cumulative frequency distributions, calculate the D statistic and give the probability.

The most common use of the Kolmogorov–Smirnov one-sample test is to compare the distribution of one sample with an expected distribution such as the normal curve, and most statistical packages include this option.

19.3 Non-parametric comparisons between two independent samples

19.3.1 The Mann–Whitney test

The Mann–Whitney test is used to compare two independent samples. Table 19.2 gives data for the length, in cm, of Devonian-age *Paracyclas* clams in each of two outcrops. The null hypothesis is that the two samples have come from the same population.

First, the values are **ranked over both samples** as shown in Table 19.2. The smallest value is given the rank of 1, the next largest the rank of 2 etc., so the largest will have the rank of $n_1 + n_2$ (which is the sum of the number of cases in each sample). For the data in Table 19.2, the largest possible rank is 9.

Table 19.2 The length, in centimeters, of Devonian-age *Paracyclas* clams in each of two outcrops. Ranks are shown in the two right-hand columns, together with the rank sums (R_1 and R_2) for each treatment.

Length in outcrop A	Length in outcrop B	Rank for outcrop A	Rank for outcrop B
24	22	7	6
41	6	9	2
17	11	5	3
38	15	8	4
	4		1
$n_1 = 4$	$n_2 = 5$	$R_1 = 29$	$R_2 = 16$

If two or more values are the same (that is, they are **tied**), each is given the average of the ranks assigned to that many values. For example, if the data in Table 19.2 contained two 4 cm long clams and these were the smallest, each would be given the average of ranks 1 and 2, which is 1.5.

If most of the clams were longer in one outcrop than the other, the ranks would differ between these two samples. In contrast, if the clams were of a similar length in both outcrops the ranks within each sample would also be similar.

The ranks are summed separately for each sample (these are R_1 and R_2 in Table 19.2) and the two Mann–Whitney statistics U and U' calculated:

$$U = n_1 \times n_2 + \frac{n_1(n_1 + 1)}{2} - R_1 \qquad (19.2)$$

and

$$U' = n_1 \times n_2 + \frac{n_1(n_2 + 1)}{2} - R_2 \qquad (19.3)$$

where n_1 and n_2 are the size of each sample.

These equations may appear complex, but are easily explained by separating them into three components as shown for U in Equation (19.2).

$$U = \underset{\text{(component A)}}{n_1 \times n_2} + \underset{\text{(component B)}}{\frac{n_1(n_1 + 1)}{2}} - \underset{\text{(component C)}}{R_1} \qquad (19.4)$$

Component A will increase with the size of both samples. Component B will only increase as the size of sample 1 increases. In contrast, component C will be affected by the way the ranks are distributed between the two samples. A lot of low ranks in sample 1 will give a relatively small value of R_1 and vice versa. Therefore, since U is calculated by taking component C away from the sum of components A and B, it will be large compared to U' when sample 1 contains mainly low ranks. In contrast, if sample 1 contains mainly high ranks, the value of U will be small compared to U'. Finally, if both samples contain similar ranks then neither U nor U' will be relatively large or small.

When both samples are from the same population, most values of U and U' will be similar but differences between them will occur by chance and the most extreme 5% of discrepancies will give values of U or U' that will exceed a critical value. For a two-tailed test, if **either** of the U statistics exceeds the critical value then the probability the samples are from the same population is less than 5%.

19.3.2　Randomization tests for two independent samples

Another way of comparing two independent samples, without assuming they are from a normal distribution, is to use a randomization test. These tests were first discussed in relation to samples of categorical data in Chapter 18.

If two independent samples are taken from the same population, then the values within each should differ only by chance. A randomization test takes the combined set of ranks from both samples (a group of size $n_1 + n_2$), repeatedly samples it at random and assigns the ranks to two groups of size n_1 and n_2.

The simulated sampling is iterated several thousand times and used to generate the **expected** distribution of U and U' from the data set and therefore identify the most extreme 5% of departures from the outcome expected under the null hypothesis. Finally, the U statistics for the actual outcome are compared to these distributions and if the probability is less than 5% it is statistically significant (Figure 19.1).

19.3.3　Exact tests for two independent samples

Data for two samples can also be analyzed by tests that calculate the exact probability, and work in a very similar way to the Fisher Exact Probability Test described in Chapter 18.

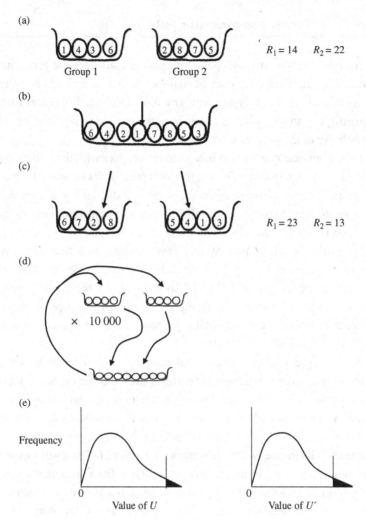

(a)

Group 1   Group 2 $R_1 = 14$ $R_2 = 22$

(b)

(c)

$R_1 = 23$ $R_2 = 13$

(d)

× 10 000

(e)

Frequency

0
Value of U

0
Value of U'

Figure 19.1 Illustration of a randomization procedure that gives distributions of the two Mann–Whitney statistics U and U' from simulated sampling, which can be used to decide whether an observed outcome is statistically significant. (a) The actual outcome of the experiment. (b) The ranks from both groups are combined. (c) The combined set of ranks is resampled at random to give two more groups of size n_1 and n_2 and thereby generate two new values of R_1 and R_2. (d) Steps (b) and (c) are repeated several thousand times. Each time, two more values of R_1 and R_2 are generated. (e) The simulated sampling gives the distributions of U and U' for two samples taken at random from the same group. By chance there will often be differences between samples, and as they increase so will U or U'. The largest 5% of the values of U and U' are shown as the filled areas on the right of each graph. Finally, the U statistics from the actual outcome (a) are compared to these distributions. If the probability of getting either U or U' is less than 5% (i.e. either statistic falls within the filled area), the null hypothesis that the samples in (a) are from the same population is rejected.

An exact test for two independent samples calculates the probability of the actual difference (or values of statistics such as U and U') between the ranks of the samples, together with any more extreme differences from the outcome expected under the null hypothesis. This gives the one-tailed probability of the outcome.

Here is an example for two independent samples with three data in each. The values range from 1 to 6, and the total set of ways in which they can be distributed between two samples is shown in Table 19.3. We have deliberately made the values the same as their ranks, and used a simple comparison between the rank sums of the samples.

For this example there are only two combinations that will give the greatest difference between the rank sums. These are when the first sample contains the three lowest (1, 2 and 3) and the second the three highest (4, 5 and 6) ranks and vice versa, giving an absolute difference of nine. Less extreme differences can be obtained from several combinations and are therefore more likely (Table 19.3).

For example, you may wish to calculate the probability of the observed outcome and any more extreme departures from the one expected under the null hypothesis when one sample contains the ranks 1, 2 and 5 (and the other contains ranks 3, 4 and 6). The observed difference between the sums of the ranks is –5. You will find this outcome in the third line from the top of Table 19.3. There are two more extreme differences (–7 and –9) in the same direction (that is, with increasingly negative values) from the outcome expected under the null hypothesis. The probability of each outcome is calculated directly from sampling a set of six values without replacement (e.g. the chance of rank 1 is 1/6, but the chance of then selecting rank 2 is 1/5 etc.). Once calculated these probabilities are summed, thus giving the one-tailed probability of the observed outcome and any more extreme departures from the null hypothesis. It is one-tailed because differences in the same absolute size between the samples (i.e. the last three lines showing differences of 5, 7 and 9) in Table 19.3 have been ignored, and has to be doubled to get the two-tailed probability.

19.3.4   Recommended non-parametric tests for two independent samples

Most statistical packages include the Mann–Whitney test, but if you have one that includes an exact test or randomization test for two

Table 19.3 The set of ways in which six ranks can be distributed between two samples of three. Note that the most extreme differences (of −9 and 9) between the sums of the ranks of two samples can only be obtained when one contains ranks 1, 2 and 3 and the other 4, 5 and 6, so these outcomes have a relatively low probability compared to less extreme differences (e.g. 1 and −1), which can be obtained in several different ways.

Sample 1			Rank sums and their differences			Sample 2		
(a)	(b)	(c)				(c)	(b)	(a)
1								4
2			$R_1 = 6$	$R_1 - R_2 = -9$	$R_2 = 15$			5
3								6
1								3
2			$R_1 = 7$	$R_1 - R_2 = -7$	$R_2 = 14$			5
4								6
1	1						2	3
2	3		$R_1 = 8$	$R_1 - R_2 = -5$	$R_2 = 13$		5	4
5	4						6	6
2	1	1				3	2	1
3	3	2	$R_1 = 9$	$R_1 - R_2 = -3$	$R_2 = 12$	4	4	5
4	5	6				5	6	6
2	1	1				2	2	1
3	3	4	$R_1 = 10$	$R_1 - R_2 = -1$	$R_2 = 11$	3	4	4
5	6	5				6	5	6
1	2	2				1	1	2
4	4	3	$R_1 = 11$	$R_1 - R_2 = 1$	$R_2 = 10$	2	3	3
6	5	6				5	6	5
1	2	3				1	1	2
5	4	4	$R_1 = 12$	$R_1 - R_2 = 3$	$R_2 = 9$	2	3	3
6	6	5				6	5	4
3	2						1	1
4	5		$R_1 = 13$	$R_1 - R_2 = 5$	$R_2 = 8$		3	2
6	6						4	5
3								1
5			$R_1 = 14$	$R_1 - R_2 = 7$	$R_2 = 7$			2
6								4
4								1
5			$R_1 = 15$	$R_1 - R_2 = 9$	$R_2 = 6$			2
6								3

Table 19.4 The length in centimeters, of Devonian-age *Paracyclas* clams from three outcrops. The totals are the rank sums within each group.

Length of clams (centimeters)			Length of clams ranked from the smallest to the largest		
Outcrop J	Outcrop K	Outcrop L	Outcrop J	Outcrop K	Outcrop L
25	31	22	9	13	8
14	20	4	4	7	1
35	29	11	14	11	3
41	15	18	16	5	6
28	40	8	10	15	2
	30			12	
		Total	$R_1 = 53$	$R_2 = 63$	$R_3 = 20$

independent samples, they are recommended in preference to the Mann–Whitney test.

19.4 Non-parametric comparisons among more than two independent samples

The most frequently used non-parametric test for more than two independent samples is the Kruskal–Wallis test. It is also called the Kruskal–Wallis single-factor analysis of variance by ranks, but this is misleading because it does not use analysis of variance to compare samples. Instead, the Kruskal–Wallis test is an extension of the Mann–Whitney test that can be applied to three or more samples.

19.4.1 The Kruskal–Wallis test

For a Kruskal–Wallis test the data are ranked in the same way as for a Mann–Whitney test, starting by assigning the lowest rank to the smallest value. Here is an example that also uses the length of Devonian-age *Paracyclas* clams discussed in Section 19.3.1, except that here the clams are being compared among three outcrops. The null hypothesis is, "There is no difference in the length of clams from the three outcrops." It is clear that a marked difference in the length of clams among outcrops will also result in a difference in the ranks and rank sums.

The rank sums for each group are used in the following formula for the Kruskal–Wallis statistic H:

$$H = \frac{12}{N(N+1)} \sum_{i=1}^{k} R_i^2 - 3(N+1) \qquad (19.5)$$

where N is the total sample size and k is the number of groups or samples. Although this formula looks complex, it is straightforward when considered as three components:

$$H = \underset{\text{(component A)}}{\frac{12}{N(N+1)}} \times \underset{\text{(component B)}}{\sum_{i=1}^{k} R_i^2} - \underset{\text{(component C)}}{3(N+1)} \qquad (19.6)$$

Components A and C will increase as sample size increases. Component B is the sum of all the squared rank totals. If all R_i values are relatively similar then component B (and therefore H) will be smaller than when some are large and others small because of the effect of squaring relatively large numbers (Box 19.1).

The distribution of H for samples taken at random from the same population has been established and used to identify the 5% most extreme departures from the null hypothesis of no difference. For large samples, or where the number of groups or treatments is more than five, the value of H is a close approximation to the chi-square statistic with $(k - 1)$ degrees of freedom, and many statistical packages only give this statistic (and its probability) for the result of a Kruskal–Wallis test.

19.4.2   Exact tests and randomization tests for three or more independent samples

Randomization and exact tests on the ranks of three or more independent samples are extensions of the methods described for two independent samples and it is not necessary to explain these further.

19.4.3   A posteriori comparisons after a non-parametric comparison

A non-parametric comparison can detect a significant difference among three or more groups, but it cannot show **which** groups appear to be from

---

Box 19.1 The effect of an unequal allocation of ranks on the total of the squared rank sums

This example uses three groups with two values in each. Only the ranks of the values are shown. First, the rank sums are identical among groups.

Group A	Group B	Group C
1	2	3
6	5	4
$R_1 = 7$	$R_2 = 7$	$R_3 = 7$

$$\sum_{i=1}^{k} R_i^2 = 3 \times 49 = 147$$

Second, the rank sums are different among groups and this gives a larger sum of the squared rank sums.

Group A	Group B	Group C
1	3	5
2	4	6
$R_1 = 3$	$R_2 = 7$	$R_3 = 11$

$$\sum_{i=1}^{k} R_i^2 = 3^2 + 7^2 + 11^2 = 179$$

---

the same, or different, populations. This problem was discussed in Chapter 10 in relation to a single-factor parametric ANOVA. If the effect of the variable you are examining is considered fixed you need to use non-parametric a posteriori tests to compare among groups. These are described in more advanced texts (e.g. Sprent, 1993).

19.4.4 Rank transformation followed by single-factor ANOVA

Another way of analyzing data that are grossly non-normal is to run a parametric single-factor ANOVA on the ranks. This is not a true non-parametric test, but has the advantage of easy a posteriori comparisons when an effect is fixed and the initial analysis shows a significant difference among samples. It is as powerful as applying a Kruskal–Wallis test.

19.4.5   Recommended non-parametric tests for three or more independent samples

Most statistical packages include the Kruskal–Wallis test, which is up to 95% as powerful as the equivalent parametric single-factor ANOVA described in Chapter 10. If you have a package that includes an exact test or randomization test, these are recommended in preference to the Kruskal–Wallis test. Several texts recommend using a parametric ANOVA after rank transformation but it is important to note that this is not a true non-parametric comparison.

19.5   Non-parametric comparisons of two related samples

Related samples were first discussed in Chapter 8. Some examples are when a variable is measured twice (and usually under different conditions) on the same sampling or experimental unit, or when the units within one sample or treatment are somehow related to those in a second (e.g. an experiment where several specimens of the same mineral are split into two, with one piece of each specimen assigned to treatment A while the other is assigned to treatment B). There are several non-parametric tests for determining the probability that two related samples have been taken from the same population and these include the Wilcoxon paired-sample test, as well as randomization and exact tests for this statistic.

19.5.1   The Wilcoxon paired-sample test

The Wilcoxon paired-sample test is the non-parametric equivalent of the paired-sample *t* test. The following example is from medical mineralogy. In Chapter 6 we mentioned that asbestos fibers show longitudinal parting and have ends that fray into individual fibers. If inhaled, these fibers often remain in the lung where they cause inflammation and scarring of lung tissue (which leads to a high incidence of cancer). In humans the trachea (the windpipe) branches into two bronchi, which lead to different lungs. The bronchus leading to the right lung is wider, shorter and angled more closely to the vertical than the one leading to the left lung. Not surprisingly, it has been found that inhaled objects are more likely to lodge in the right bronchus, so a pathologist hypothesized the right lung may also receive a greater proportion of inhaled asbestos particles and therefore show a higher incidence of lesions.

To test this, the pathologist counted the number of lesions found during post mortem examination of the left and right lungs of 10 male asbestos workers who had died of natural causes. The data are in Table 19.5.

For the Wilcoxon test the difference between each pair of related sampling units is first calculated. Each is also given as the absolute difference and these values are ranked (Table 19.5). Finally, the ranks associated with negative and positive differences are summed separately to give the Wilcoxon statistics $T+$ and $T-$. For the data in Table 19.5, the ranks of the positive differences sum to 25 (cases 1, 2, 4, 5, 6 and 9) while the ranks of the negative differences sum to 30 (cases 3, 7, 8 and 10).

Under the null hypothesis of no effect of bronchial structure on the number of lesions in each lung, any differences between each pair of related samples (and therefore $T+$ and $T-$) would only be expected by chance. If, however, there were an effect of bronchial structure it would contribute to differences between these two statistics.

The values of $T+$ and $T-$ can be compared to their expected distributions from taking related samples at random from a population. For a two-tailed test the null hypothesis is rejected if **either** $T+$ or $T-$ is **less** than a critical value, but for a one-tailed test the null hypothesis is only rejected if the appropriate T statistic is less than a critical value. For example, if it were hypothesized there were more lesions in the right lung than the left, a reduction in the number of negative ranks would be expected so the null hypothesis would only be rejected if $T-$ were less than the critical value.

For large samples the distributions of both T statistics approximate the normal curve, so statistical packages often give the value of the Z statistic and probability for the result of the Wilcoxon test.

19.5.2 Exact tests and randomization tests for two related samples

The procedures for randomization and exact tests on the ranks of two related samples are conceptually similar to the analyses for two independent

Table 19.5 Data for the number of separate lesions found during post mortem examination of the left and right lungs of 10 male asbestos workers who died from natural causes. For a Wilcoxon test, the difference between each pair of related data is calculated (d), expressed as the absolute difference (e) and these absolute values ranked (f). The ranks associated with positive differences (h) and negative differences (i) are separately summed to give the statistics $T+$ and $T-$.

(a) Specimen number	(b) Right lung	(c) Left lung	(d) Difference (right − left)	(e) Absolute difference	(f) Rank of the absolute difference	(g) Sign of the difference	(h) Ranks associated with positive differences	(i) Ranks associated with negative differences
1	19	15	4	4	2	+	2	
2	24	14	10	10	6	+	6	
3	16	22	−6	6	4	−		4
4	28	28	0	0	1	+	1	
5	19	11	8	8	5	+	5	
6	26	9	17	17	8	+	8	
7	16	38	−22	22	10	−		10
8	27	42	−15	15	7	−		7
9	18	13	5	5	3	+	3	
10	18	37	−19	19	9	−		9
						Totals	$T+ = 25$	$T- = 30$

samples described in Section 19.3 and it is not necessary to explain them any further.

19.6   Non-parametric comparisons among three or more related samples

Tests for three or more related samples include the Friedman test, together with randomization and exact tests for this statistic.

19.6.1   The Friedman test

The Friedman test is often called the Friedman two-way analysis of variance by ranks, but this is misleading because it is not equivalent to the two-factor ANOVA discussed in Chapter 12. The Friedman test cannot detect inter-action and only examines differences among the levels of one factor, so is really analogous to the two-factor ANOVA without replication applied to the randomized block experimental design described in Chapter 14.

Table 19.6 gives the results of a randomized block experiment designed to compare the effects of the addition of vermiculite (Section 8.4.3) upon the amount of water retained by topsoil in a large field. There is consid-erable natural spatial variation in water retention, so the experimental field was subdivided into six strips and one replicate of every treatment

Table 19.6 The number of grams of water per 100 grams of topsoil three weeks after one replicate of two vermiculite treatments and a control were applied to each of six blocks in a large field.

Block	Treatment A: 10 g/kg vermiculite	Treatment B: 50 g/kg vermiculite	Control (plowed only)	Rank of Treatment A	Rank of Treatment B	Rank of control
1	2.5	2.7	2.1	2	3	1
2	1.8	1.9	2.0	1	2	3
3	4.4	4.7	4.1	2	3	1
4	2.4	2.6	2.3	2	3	1
5	5.1	5.3	5.2	1	3	2
6	1.7	1.9	1.6	2	3	1
			Totals	$R_1 = 10$	$R_2 = 17$	$R_3 = 9$

assigned to each strip, in a randomized block design with three treatments and six blocks. Soil in the control treatment was plowed, while soil in the other two treatments was (a) plowed and mixed with 10 g/kg vermiculite and (b) 50 g/kg vermiculite. Data for the number of grams of water per 100 grams of topsoil three weeks after the treatments were applied are in Table 19.6.

For a Friedman test the data are first transformed to ranks. These are assigned **within each block** and therefore within **each row** of Table 19.6. The lowest value in each row is given the rank of "1", the next highest "2" etc., and the highest rank cannot exceed the number of treatments.

If the treatments are from the same population, the range of ranks (and the rank sums) for each should also be similar, with any variation due to chance. If, however, there is any effect of either treatment, the ranks and their sums will also differ. For the example in Table 19.6, the control contains all but two of the lowest ranks, while treatment B (50 g/kg vermiculite) contains all but one of the highest.

Next, the total of the squared rank sums is calculated. The size of this total will depend on the relative size of the rank sums (Box 19.1) with a set of similar ones giving a smaller total than a set of dissimilar ones.

Finally the following formula is used to calculate the Friedman statistic χ_r^2:

$$\chi_r^2 = \frac{12}{ba(a+1)} \sum_{i=1}^{a} R_i^2 - 3b(a+1) \tag{19.7}$$

where a is the number of treatments or groups and b is the number of blocks. This appears complex, but can be split into three components as shown in equation (19.8) below. The Friedman statistic is obtained by multiplying components A and B together and then subtracting component C.

$$\chi_r^2 = \underbrace{\frac{12}{ba(a+1)}}_{\text{(component A)}} \times \underbrace{\sum_{i=1}^{a} R_i^2}_{\text{(component B)}} - \underbrace{3b(a+1)}_{\text{(component C)}} \tag{19.8}$$

Components A and C will increase as sample sizes and the number of samples increase. If the rank sums are very similar among treatments, component B will be relatively small so the value of the Friedman statistic will also be small. As the differences among the rank sums increase, component B will increase, thus giving an increasingly larger value of the

Friedman statistic. Once this exceeds the critical value above which less than 5% of the most extreme departures from the null hypothesis occur when samples are taken from the same population, the outcome is considered statistically significant.

This analysis can be up to 95% as powerful as the equivalent two-factor ANOVA without replication for randomized blocks.

19.6.2 Exact tests and randomization tests for three or more related samples

The procedures for randomization and exact tests on the ranks of three or more related samples are extensions of the methods for two independent samples and do not need to be explained any further.

19.6.3 A posteriori comparisons for three or more related samples

If the Friedman test shows a significant difference among treatments and the effect is considered fixed, you are likely to want to know which treatments are significantly different (see 19.4.3). A posteriori testing can be done and instructions are given in more advanced texts such as Zar (1996).

19.7 Analyzing ratio, interval or ordinal data that show gross differences in variance among treatments and cannot be satisfactorily transformed

Some data show gross differences in variance among treatments that **cannot** be improved by transformation and are therefore unsuitable for parametric or non-parametric analysis. An exploration geologist in Canada was evaluating the economic potential of a circular depression thought to be an impact crater. They knew that the large-scale impact structure at nearby Sudbury was associated with valuable copper and nickel deposits, and that other impact structures are excellent reservoirs for oil and gas. So they set out to determine if the new locality might also be an impact structure.

One of the key properties of impacted rocks is their high concentration of platinum group elements. Perhaps the most diagnostic of these is iridium, which is famously found all over the world in an ash layer that corresponds to the end of the Cretaceous Period and the extinction of the dinosaurs. Iridium is not normally present in crustal rocks on the Earth's surface – it is usually found only in the metallic cores of differentiated planets and in iron from meteorites. So when an impact from an iron-rich object occurs on Earth, the iridium vaporizes and is distributed among the impact ejecta in unusually high concentrations (up to 100 parts per billion). Thus iridium concentration can be used as a geochemical tracer to indicate that rocks have experienced an impact event.

The exploration geologist collected 15 core samples from his suspected new impact site, along with 15 from the Sudbury impact structure. The concentration of iridium in the two samples of 15 is given in Table 19.7.

It is clear there are gross differences in the distributions between the two samples, with one showing bimodality. A solution is to transform the data to a nominal scale and reclassify both samples into two mutually exclusive categories of "with iridium" and "no iridium" (Table 19.8) which can be compared using a test for two or more independent samples of categorical data (Chapter 18).

Table 19.7 The Ir contents (in parts per billion) of 15 rocks sampled at Sudbury crater (a classic impact site) and 15 at a new site with a circular feature suspected to be an impact crater.

Sudbury	New site
4	2
7	0
4	2
10	0
2	0
7	0
1	0
9	0
1	1
9	0
12	1
1	0
5	0
4	1
5	0

Table 19.8 Transformation of the ratio data in Table 19.7 to a nominal scale showing the number of replicates in each sample as the two mutually exclusive categories of with and without detectable iridium.

	Sudbury	New site
Number without iridium	0	10
Number with iridium	15	5

19.8   Non-parametric correlation analysis

Correlation analysis was introduced in Chapter 15 as an exploratory technique used to examine whether two variables are related or **vary together**. Importantly, there is no expectation that the numerical value of one variable can be predicted from the other, nor is it necessary that either variable is determined by the other.

The parametric test for correlation gives a statistic that varies between +1.00 and –1.00, with both of these extremes indicating a perfect positive and negative straight line relationship respectively, while values around zero show no relationship. Although parametric correlation analysis is powerful, it can only detect linear relationships and also assumes that both the X and Y variables are normally distributed. When normality of both variables cannot be assumed, or the relationship between the two variables does not appear to be linear and cannot be remedied by transformation, it is not appropriate to use a parametric test for correlation. The most commonly used non-parametric test for correlation is Spearman's rank correlation.

19.8.1   Spearman's rank correlation

This test is extremely straightforward. The two variables are ranked separately, from lowest to highest, and the (parametric) Pearson correlation coefficient calculated for the ranked values. This gives a statistic called Spearman's rho, which for a population is symbolized by ρ_s and by r_s for a sample.

Spearman's r_s and Pearson's r will not always be the same for the same set of data. For Pearson's r the correlation coefficients of 1.00 or –1.00 were

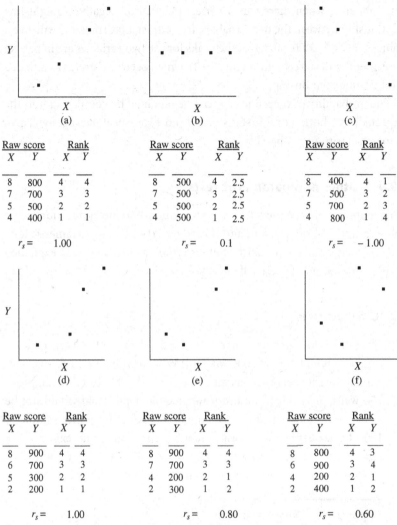

Figure 19.2 Examples of raw scores, ranks and the Spearman rank correlation coefficient for data with: (a) a perfect positive relationship (all points lie along a straight line); (b) no relationship; (c) a perfect negative relationship (all points lie along a straight line); (d) a positive relationship which is not a straight line but all pairs of bivariate data have the same ranks; (e) a positive relationship with only half the pairs of bivariate data having equal ranks; (f) a positive relationship with no pairs of bivariate data having equal ranks. Note that the value of r_s is 1.00 for case (d) even though the raw data do not show a straight-line relationship.

only obtained when there was a perfect positive or negative straight-line relationship between the two variables. In contrast, Spearman's r_s will give a value of 1.00 or -1.00 whenever the ranks for the two variables are in perfect agreement or disagreement (Figure 19.2), which occurs in more cases than a straight-line relationship.

The probability of the value of r_s can be obtained by comparing it to the expected distribution of this statistic and most statistical packages will give r_s together with its probability.

19.9　Other non-parametric tests

This chapter is only an introduction to some non-parametric tests for two or more samples of independent and related data. Other non-parametric tests are described in more specialized but nevertheless extremely well-explained texts such as Siegel and Castallan (1988).

19.10　Questions

(1) The table below gives summary data for the depth of the water table, in feet, for a population of 1000 wells. (a) What are the relative frequencies and cumulative relative frequencies for each depth? (b) For a sample of 100 wells, give a distribution of water table depths that would not be significantly different from the population. (c) For another sample of 100 give a distribution of water table depths you would expect to be significantly deeper than the population. (d) What test would be appropriate to compare these samples to the known population?

Depth (feet)	Number of wells
20–29	150
30–39	300
40–49	140
50–59	110
60–69	30
70–79	110
80–89	140
90–99	20

(2) An easy way to understand the process of ranking, and the tests that use this procedure, is to use a contrived data set. The following two independent samples have very similar rank sums. (a) Rank the data across both samples and calculate the rank sums. (b) Use a statistical package to run a Mann–Whitney test on the data. Is there a significant difference between the samples? (c) Now change the data so you would expect a significant difference between groups. Run the Mann–Whitney test again. Was the difference significant?

Group 1	Group 2
4	5
7	6
8	9
11	10
12	13
15	14
16	17
19	18
20	21

(3) The following set of data for the percentage of sandstone porosity shows a gross difference in distribution between two samples. (a) How might you compare these two samples? (b) Use your suggested method to test the hypothesis that the two samples have different porosities. Is there a significant difference?

Sample 1:   1, 1, 1, 1, 2, 1, 1, 1, 2, 1, 1, 1, 1, 1, 1, 1, 2, 2, 2, 3, 5, 2
Sample 2:   1, 1, 1, 1, 1, 1, 10, 11, 11, 11, 12, 12, 13, 13, 13, 13, 14, 14, 15, 17, 18, 18, 19

20 | Introductory concepts of multivariate analysis

20.1   Introduction

So far, all the analyses discussed in this book have been for either univariate or bivariate data. Often, however, earth scientists need to analyze samples of **multivariate** data – where **more than two variables are measured on each sampling or experimental unit** – because univariate or bivariate data do not give enough detail to realistically describe the material or the environment being investigated.

For example, a large ore body may contain several different metals, and the concentrations of each of these may vary considerably within it. It would be useful to have a good estimate of this variation because some parts of the deposit may be particularly worth mining, others may not be worth mining at all, or certain parts may have to be mined and processed in different ways. Data for only one or two metals (e.g. copper and silver) are unlikely to be sufficient to estimate the full variation in composition and value within a deposit that also includes lead and zinc.

Samples on which multivariate data have been measured are often difficult to compare with one another because there are so many variables. In contrast, samples where only univariate data are available can easily be visualized and compared (e.g. by summary statistics such as the mean and standard error). Bivariate data can be displayed on a two-dimensional graph, with one axis for each variable. Even data for three variables can be displayed in a three-dimensional graph. But as soon as you have four or more variables, the visualization of these in a multidimensional space and comparison among samples becomes increasingly difficult. For example, Table 20.1 gives data for the concentrations of five metals at four sites. Although this is only a small data set, it is difficult to assess which sites are most similar or dissimilar. (Incidentally, you may be thinking this is a

Table 20.1 The concentrations of five metals at four sites (A–D). From these raw data, it is difficult to evaluate which sites are most similar or dissimilar.

Metal	Site A	Site B	Site C	Site D
Copper	12	43	26	21
Silver	11	40	28	19
Lead	46	63	26	21
Gold	32	5	19	7
Zinc	6	40	21	38

very poor sampling design, because data are only given for one sampling unit at each site. This is true, but here we are presenting a simplified data set for clarity.)

Earth scientists need **ways of simplifying and summarizing multivariate data** to compare samples. Because univariate data are so easy to visualize, the comparison among the four sites in Table 20.1 would be greatly simplified if the data for the five metals could somehow be **reduced to a single statistic or measure**. Multivariate methods do this by reducing the complexity of the data sets while retaining as much information as possible about each sample. The following explanations are simplified and conceptual, but they do describe how these methods work.

20.2   Simplifying and summarizing multivariate data

The methods for simplifying and comparing samples of multivariate data can be divided into two groups.

(a) The first group of analyses works on the variables themselves. They **reduce the number of variables** by identifying **the ones that have the most influence upon the observed differences among sampling units** so that **relationships among the units** can be summarized and visualized more easily. These "variable-oriented" methods are often called **R- mode analyses**.

(b) The second group of analyses works on the sampling units. They often summarize the multivariate data by calculating **a single measure, or statistic**, that helps to **quantify differences among sampling units**. These "sample-oriented" methods are often called **Q-mode analyses**.

This chapter will describe an example of an R-mode analysis, followed by two Q-mode ones.

20.3   An R-mode analysis: principal components analysis

Principal components analysis (PCA) (which is called "principal component analysis" in some texts) is one of the oldest multivariate techniques. The mathematical procedure of PCA is complex and uses matrix algebra, but the concept of how PCA works is very easy to understand. The following explanation only assumes an understanding of the correlation between two variables (Chapter 15).

If you have a set of data where you have measured several variables on a set of sampling units (e.g. a number of sites or cores), which for PCA are often called **objects**, it is very difficult to compare them when you have data for more than three variables (e.g. the data in Table 20.1).

Quite often, however, a set of multivariate data shows a lot of **redundancy** – that is, two or more variables are **highly correlated** with each other. For example, if you look at the data in Table 20.1, it is apparent that the concentrations of copper, silver and zinc are positively correlated (when there are relatively high concentrations of copper there are also relatively high concentrations of silver and zinc and vice versa). Furthermore, the concentrations of copper, silver and zinc are also correlated with gold, but we have deliberately made these correlations negative (when there are relatively high concentrations of gold, there are relatively low concentrations of copper, silver and zinc and vice versa) because negative correlations are just as important as positive ones.

These correlations are an example of **redundancy** within the data set – because four of the five variables are well-correlated, and knowing which correlations are negative and which are positive, **you really only need the data for one of these variables** to describe differences among the sites. Therefore, you could reduce the data for these four metals down to only one (copper, silver, gold or zinc) plus lead in Table 20.2 with little loss of information about the sites.

A principal components analysis uses such cases of redundancy to reduce the number of variables in a data set, although it does not exclude variables. Instead, PCA **identifies variables that are highly correlated** with each other and combines these to **construct a reduced set of new variables that still**

Table 20.2 Because the concentrations of copper, silver, gold and zinc are correlated, you only need data for one of these (e.g. silver), plus the concentration of lead, to describe the differences among the sites.

Metal	Site A	Site B	Site C	Site D
Silver	11	40	28	19
Lead	46	63	26	21

describes the differences among samples. These new variables are called **principal components** and are listed in decreasing order of importance (beginning with the one that explains the most variation among sampling units, followed by the next greatest, etc.). With a reduced number of variables, any differences among sampling units are likely to be easier to visualize.

20.4   How does a PCA combine two or more variables into one?

This is a straightforward example where data for two variables are combined into one new variable, and we are using a simplified version of the conceptual explanation presented by Davis (2002). Imagine you need to assess variation within a large ore body for which you have data for the concentration of silver and gold at ten sites. It would be helpful to know which sites were most similar (and dissimilar) and how the concentrations of silver and gold varied among them.

The data for the ten sites have been plotted in Figure 20.1, which shows a negative correlation between the concentrations of silver and gold. This strong **relationship between two variables** can be used to **construct a single, combined variable to help make comparisons among the ten sites**. Note that you are not interested in whether the variables are positively or negatively correlated – you only want to compare the sites.

The bivariate distribution of points for these two highly correlated variables could be enclosed by a **boundary**. This is analogous to the way a set of univariate data has a 95% confidence interval (Chapter 8). For this bivariate data set the boundary will be two dimensional, and because the variables are correlated it will be elliptical as shown in Figure 20.2.

An ellipse is symmetrical and its relative length and width can be described by the length of the longest line that can be drawn through it

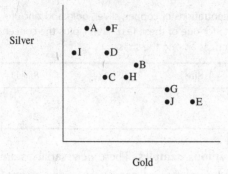

Figure 20.1 The concentration of silver versus the concentration of gold at ten sites.

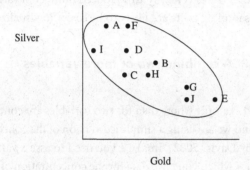

Figure 20.2 An ellipse drawn around the set of data for the concentration of silver versus the concentration of gold in ore at ten sites. The elliptical boundary can be thought of as analogous to the 95% confidence interval for this bivariate distribution.

(which is called the major axis), and the length of a line drawn halfway down and perpendicular to the major axis (which is called the minor axis) (Figure 20.3).

The relative lengths of the two axes describing the ellipse **will depend upon the strength of the correlation between the two variables**. Highly correlated data like those in Figure 20.3 will be enclosed by a long and narrow ellipse, but for weakly correlated data the ellipse will be far more circular.

At present the ten sites are described by two variables – the concentrations of silver and gold. But because these two variables are highly correlated, all the sites are quite close to the major axis of the ellipse, so most of the variation among them can be described by just that axis (Figure 20.3). Therefore, you can think of the major axis as a **new single variable** that is a good indication of

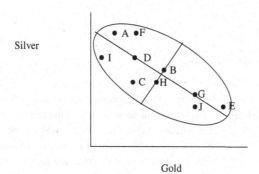

Silver

Gold

Figure 20.3 The long major axis and shorter minor axis give the dimensions of the ellipse that encloses the set of data.

most of the variation among sites. **So instead of using two variables to describe the ten sites, the information can be combined into just one.**

The two axes are called **eigenvectors** and the relative length of each that falls within the ellipse is its **eigenvalue.** Once the longest eigenvector of the ellipse has been drawn, it is rotated (in the case of Figure 20.3 this will simply be anticlockwise by about 45°) so that it becomes the new X axis (Figure 20.4). This new, **artificially constructed principal component** explains most of the variation among the ten sites. It has no name except principal component number 1 (PC1). It is important to remember that PC1 is a new variable – in this case it is a **combination** of the two variables "concentration of silver" and "concentration of gold." The plot of the points in relation to PC1 in Figure 20.4 only shows the sites in terms of this new variable – there is nothing about silver or gold in the graph.

The new X axis, PC1, is rescaled to assign the midpoint of the axis the value of zero. This makes the axis symmetrical about zero, so the objects will have both positive and negative coordinates for PC1 (Figure 20.5).

In this example, the points are all close to the major axis, so principal component 1 explains the majority of the variation among the sites, and can be used to easily assess similarities among them. From Figures 20.4 and 20.5 it is clear that sites A, I and F are more similar to each other than A is to E because the distance between the former three is much shorter.

Because there are two variables in the initial data set, principal components analysis also constructs a second component that is completely independent and uncorrelated with principal component 1. The second axis is called principal component 2 (PC2) and is simply the minor axis of the ellipse

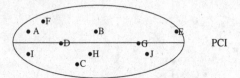

Figure 20.4 The long axis of the ellipse has been drawn through the set of highly correlated data for the concentration of silver and the concentration of gold (Figure 20.3), and then rotated to give a new X axis (which is the major axis of the ellipse) for the artificial variable called principal component number 1. This new variable explains most of the variation among sites.

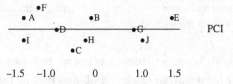

Figure 20.5 The values for PC1 are expressed in relation to the midpoint of the principal eigenvector, which is assigned the value of zero.

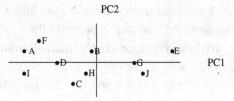

Figure 20.6 Principal component 2 is the short axis of the ellipse shown in Figure 20.5 and constructed by drawing a line perpendicular to the line showing PC1. Note that PC2 explains very little of the variation among sites.

shown in Figure 20.3, which after the rotation described above will be a line perpendicular to PC1. Here too, the eigenvalue for PC2 corresponds to its relative length and its midpoint is given the value of zero. It is clear that PC2 does not explain very much of the variation among the sites – the objects are quite widely dispersed around it, so it is a relatively short eigenvector (Figure 20.6). Therefore, most of the variation is described by PC1, and the analysis has effectively reduced the number of variables from two to one.

20.5   What happens if the variables are not highly correlated?

As described above, if the two variables **are highly correlated** the ellipse enclosing the data will be very long and narrow. Therefore the first

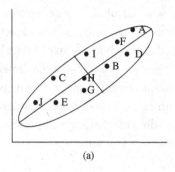

(a)

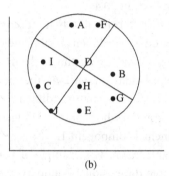

(b)

Figure 20.7 (a) Highly correlated data. The long axis is a good indication of variation among sites. (b) Uncorrelated data. The major and minor axes of the ellipse surrounding the data points are both similar in length. Therefore neither axis is a good single summary of the variation among sites.

eigenvector will be relatively long with a large eigenvalue, and the second will be relatively short with a small eigenvalue. In this case, **by itself the new combined variable of the first eigenvector is a good indicator of the differences among sites.**

In contrast, if the two variables **are not correlated** the ellipse will be more circular and the first and second eigenvectors will both have similar eigenvalues (Figure 20.7). Therefore, **neither can be used by themselves as a good indication of the differences among sites.**

20.6 PCA for more than two variables

Principal components analysis becomes particularly useful when you have data for three or more variables.

If you have n variables a PCA will calculate n eigenvectors (with n eigenvalues) that give the dimensions of an n-dimensional object in an n-dimensional space. This may sound daunting but it is easy to visualize for only three variables, where the three eigenvectors will give the dimensions for a three-dimensional object in three-dimensional space. The object will be close to spherical for a data set with no correlations and therefore little redundancy, but a very elongated three-dimensional hyperellipsoid for a set of two or three highly correlated variables. The same applies to however many additional dimensions there are.

For three or more variables the PCA procedure is an extension of the explanation given for two variables in Section 20.4.

The longest axis of the object is found and rotated so that it becomes the X axis lying horizontally to the viewer on a two-dimensional plane with its flat surface facing the viewer (like the page you are reading at the moment). If there are many variables and therefore many dimensions, the rotation is likely to be complex – for example, an eigenvector in three dimensions may have to be rotated in both the transverse and the horizontal. The eigenvector for the longest axis then becomes principal component 1.

After this the other eigenvectors are drawn. For example, if you have measured three variables, then the three-dimensional boundary enclosing the data points will have three eigenvectors describing its length, breadth and depth, all at 90° to each other.

In many cases **several** variables may be highly correlated with each other, so the hyperellipsoid may be relatively simple and may even describe most of the variation among sites in just one or two dimensions.

Here is an example. An environmental geochemist sampled sediments along a 100 mile section of coastline, including five estuaries (A–E) that received storm water runoff from urban areas and five control estuaries (F–J) that did not. At each site, they obtained data for the concentration of copper, lead, chromium, nickel, cadmium, aluminum, mercury, zinc, total polycyclic aromatic hydrocarbons (ΣPAHs) and total polychlorinated biphenyls (ΣPCBs). These ten variables were subject to principal components analysis and re-expressed as ten principal components giving the shape of a ten-dimensional hyperellipsoid. Because several of the initial variables were highly correlated, the first principal component (PC1) **explained 70% of the variation among estuaries**. The second, PC2, explained 15% more of the variation and the third, PC3, only 5% of the

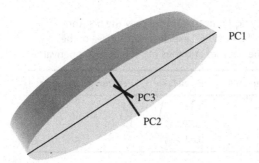

Figure 20.8 Because several variables are highly correlated they can be re-expressed as a hyperellipsoid with one very long axis (PC1), a shorter one (PC2) and a very short one (PC3). Most of the variation can be explained by PC1 and PC2. The third component, PC3, accounts for very little variation and could be ignored.

variation. Therefore, in this case 85% of the variation among site could be described by a two-dimensional ellipse with axes of PC1 and PC2, and 90% could be described by a three-dimensional ellipse with axes of PC1, PC2 and PC3. So the three-dimensional hyperellipsoid will approximate a very elongate, not very wide, and even less thick object suspended in three-dimensional space (Figure 20.8) and the remaining seven dimensions will make little contribution to its shape.

Therefore, you could take only PC1 and PC2 and plot a two-dimensional ellipse from which you can easily visualize the relationships among the sites. The two principal components explain 85% of the variation, so the closeness of the objects in two dimensions will give a realistic indication of their similarities (Figure 20.9). The analysis shows two relatively distinct clusters corresponding to the five urban and five control estuaries, consistent with urban storm water runoff having a relatively consistent effect (although you need to bear in mind that this is only a mensurative experiment).

20.7 The contribution of each variable to the principal components

Although the analysis described above has reduced the ten variables to two principal components, it is often useful to know which specific variables contribute to each of these components. For example, most of the variation (i.e. PC1) might only be related to ΣPAHs and ΣPCBs; such an outcome might suggest ways of reducing the effects of urban development upon

Table 20.3 Typical output table for only the first three components of a PCA. PC1 explains most (70%) of the variation in the data set and thus has the largest eigenvalue.

Principal component	Eigenvalue	Percentage variation
1	3.54	70
2	1.32	15
3	0.64	5

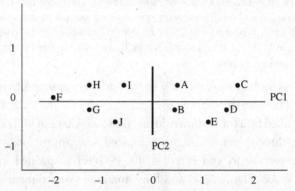

Figure 20.9 A plot of only PC1 and PC2 can still explain most of the variation among sites A–J. Note that the five urban estuaries are clustered to the right of the plot and the five control estuaries are clustered to the left.

estuaries. To address questions such as these, PCA also gives the relative contribution of each variable to each component.

The output from a PCA usually includes a plot such as Figure 20.9 and a table of **eigenvalues**. As described above, an eigenvalue gives the relative length of each eigenvector for the dimensions of the hyperellipsoid. As an example, a list of eigenvalues is given in Table 20.3, which also gives **the percentage of variation explained by each principal component**. Here too the hyperellipsoid is non-spherical, so you know the variables show redundancy and the PCA procedure has usefully reduced the number of variables.

Importantly, as well as reducing the number of variables to help visualize the relationships among objects, **PCA also gives the relative contribution of the original variables to each eigenvalue.** The output table from a PCA will contain a list of the original variables and their correlations with each of the principal components. Table 20.4 gives an example for the ten variables in the

Table 20.4 Typical output table from a PCA. The far left-hand column lists the original variables (in this case, variables 1–10) and the elements they represent. The next three columns represent the first three principal components and the values in these columns are the correlations between the new components and the original variables. Note that PC1 is primarily composed of the concentrations of variables 3 and 6 (the two largest values for the correlation coefficients and shown in bold) while PC2 is primarily composed of the concentrations of variables 1, 2 and 10 (also bold). The variables that contribute most to PC3 are 4 and 5.

Original variable	Component 1	Component 2	Component 3
1 Copper	0.01	**0.60**	0.22
2 Lead	0.24	**0.61**	0.37
3 Chromium	**0.91**	0.26	−0.06
4 Nickel	−0.18	0.32	0.57
5 Cadmium	0.15	0.05	0.52
6 Aluminum	**−0.87**	−0.22	0.44
7 Mercury	0.42	0.19	0.37
8 Zinc	0.30	−0.02	−0.22
9 ΣPAHs	−0.17	0.21	−0.06
10 ΣPCBs	0.05	**−0.71**	0.32

estuarine study described above. It is clear that principal component 1 is mainly composed of variables 3 and 6, which are chromium and aluminum (the two highest positive and negative correlations). In contrast, principal component 2 is largely composed of variables 1, 2 and 10, which are copper, lead and ΣPCBs. Which two variables make the major contribution to principal component 3? You need to look for the highest correlations, irrespective of their signs. (They are nickel and cadmium.)

The signs of the correlations are also useful. For example, for principal component 1 (Table 20.4), the correlation coefficient for variable 3 (chromium) is positive, and the one for variable 6 (aluminum) is negative. This means that as PC1 increases, chromium concentration also increases, but aluminum decreases.

In summary, a PCA has the potential to express multivariate data in a form that we can more easily understand, by reducing the number of dimensions so the data can be plotted on a two- or three-dimensional graph. It also gives a good indication of which variables contribute most to the differences among sampling units.

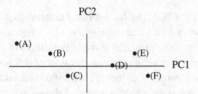

Figure 20.10 A plot of PC1 and PC2 for six sites increasingly distant (site A = closest, site F = most distant) from a petrochemical plant. The analysis shows a clear gradation through sites A to F.

20.8 An example of the practical use of principal components analysis

A marine geochemist was interested in comparing the hydrocarbons in sediments at six sampling sites, each one mile apart, running south along the shore and increasingly distant from a petrochemical plant to (a) see if there were differences in hydrocarbon levels among the sites, and (b) if so, to find out which compounds might be the best indicators of pollution.

The geochemist sampled ten hydrocarbons at each of six sites (A–F). A principal components analysis showed that only two hydrocarbons, 1 and 6 (combined as PC1), contributed to most of the variation among sites and were negatively correlated with PC1, followed by 5 and 9 (combined as PC2). When plotted on a graph of PC1 and PC2 there was a clear pattern (Figure 20.10) in that the rank order of the sites, running from left to right, corresponded to their distance from the petrochemical plant. Thus they concluded that the concentrations of only two hydrocarbons can explain most of the variation among sites.

20.9 How many principal components should you plot?

There are several ways of deciding upon how many components to use in a plot. If you are lucky, you might be in the situation where only one or two are needed, but this will only occur if they account for almost all the percentage variation among sampling units. **Generally, however, you should not use components with eigenvalues of 1.0 or less,** because this is the level of variation that you would expect by chance when there are no strong correlations among variables and therefore all original variables contribute equally to a component.

20.10 How much variation must a PCA explain before it is useful?

Very generally, if the first two or three components describe more than 70% of the variation among sampling units, then the analysis will produce a plot in two or three dimensions that is reasonably realistic.

Sometimes, however, it may be useful to know that **none** of the variables within a multivariate data set can explain very much of the variation among sampling units. For example, PCA of a multivariate data set for indicators of air pollution (nitrogen dioxide, sulfur dioxide, ozone, ammonia and the concentration of fine particles per cubic meter of air) at sites throughout a city, including the center and the fringes of the outer suburbs, showed no component with an eigenvalue greater than 0.9; none explained more than 16% of the variation among sites. The two-dimensional plot of the data was almost circular, and the three-dimensional plot was spheroidal. It was concluded that there was no obvious difference in air quality (in relation to these five indicators) across the city.

20.11 Summary and some cautions and restrictions on use of PCA

PCA is a way of reducing the complexity of a multivariate data set, but it can only do this if some variables are highly correlated. Any highly correlated variables are combined to form principal components, which may allow sampling units on which multivariate data have been measured to be plotted in two or three dimensions. The contribution of each original variable to the principal components is also given.

PCA is best suited to data where there are few zero values (e.g. grain size or concentration). It is not well suited for data such as counts, where many cells in the table of sites versus variables have a count of zero (e.g. the number of diamonds in each of several 1 m^3 sampling units of kimberlite). This restriction can be thought of in terms of the PCA constructing new axes from highly correlated variables. If the data contain a lot of zero values for each variable with only some larger numbers, the PCA is likely to overestimate redundancy, just as a group of points close to zero and a few points within a bivariate plot are likely to overestimate the strength of a correlation.

The plot provided by a PCA is also sensitive to the scale on which each variable is measured. For example, data for the concentrations of ten metals might include rare ones measured in ng/g of sediment and more abundant ones in g/kg of sediment. This will affect the shape of the hyperellipsoid, and if the data are rescaled (e.g. all expressed as ng/g) the PCA plot will stretch or shrink to reflect this. One solution, which is often automatically applied by many PCA programs, is to **normalize** the data. This is done by converting each datum to a standard Z score, as described in Chapter 8. For each variable, every datum is subtracted from the mean and the difference divided by the standard deviation. This always gives a distribution with a mean of zero and a standard deviation of 1.0, which provides a way of standardizing the data, in just the same way that a data set was standardized for a correlation analysis in Chapter 15, Equation (15.2).

20.12 Q-mode analyses: multidimensional scaling

Q-mode analyses are similar to R-mode ones in that they also reduce the effective number of variables in a data set, but they do it in a different way.

The previous sections describe how PCA combines highly correlated **variables** in order to create fewer new ones. **In contrast, multidimensional scaling (MDS) examines the similarities among sampling units.** For example, you might have data for ten variables (e.g. the concentrations of ten different hydrocarbons) measured at each of three polluted and three unpolluted sites. As discussed in relation to principal components analysis, if you were to graph all ten variables, you would need a ten-dimensional graph that would be impossibly difficult to interpret.

Multidimensional scaling is another way of condensing multivariate information so that samples can usually be displayed on a graph with fewer dimensions than the number of variables in the original data set. This method takes the data for the original set of samples and calculates a single measure of the **dissimilarity between each of the possible pairs of these**. These dissimilarity data, which are univariate, are then used to draw a plot of the samples in two- (or three-) dimensional space. Here is a very straightforward example.

Imagine that you are interested in the spatial relationships among pegmatites within a specific magmatic system. If you were to take four different pegmatites (for now we will call them A, B, C and D) within a few adjacent counties or quadrangles and measure the distances between every possible pair of these (A–B, A–C, A–D, B–C, B–D, C–D), then you could construct the matrix shown in Table 20.5. These data indicate the **dissimilarity** between

Table 20.5 The dissimilarities, expressed as distance apart in kilometers, for four pegmatites. Those close together will have a low dissimilarity score, while for those further apart the score will be higher. Note that each pegmatite is no distance from itself. The values are duplicated (i.e. the distance between Newry and Phillips is the same as that between Phillips and Newry) and the matrix is symmetrical: you only need the similarities either above or below the diagonal showing values of zero.

	Streaked Mtn.	Mount Mica	Newry	Phillips
Streaked Mtn.	0	7	58	84
Mount Mica	7	0	50	80
Newry	58	50	0	76
Phillips	84	80	76	0

pegmatites in terms of their distance apart: pegmatites that are very close together have a low score, while those further apart have a higher one.

Knowing the dissimilarity values from the matrix you could draw at least one map showing the position of the pegmatites in two dimensions. Not all of the maps would match the actual position of the pegmatites on a real geologic map, but they would be a convenient way of visualizing the relationships among the pegmatites. Two examples are shown in Figure 20.11.

This is what multidimensional scaling does. The example using pegmatites is very simple, but if you have a matrix of dissimilarities among sampling units you can use these univariate data to position the units in only two dimensions and easily visualize how closely they are related. Those close to each other will be more similar than those further apart.

20.13 How is a univariate measure of dissimilarity among sampling units extracted from multivariate data?

Univariate measures such as the **Euclidian distance** can be used to indicate dissimilarity between sampling units for which multivariate data are available. The Euclidian distance is just the distance between any two sampling units in two-, three-, four- or higher-dimensional space.

Here is an example for only two dimensions. The length of the hypotenuse of a triangle is the square root of the sum of the squared lengths of the two other sides of the triangle (Figure 20.12). For example, for two points (A and B) in two-dimensional space, with axes of Y_1 and Y_2 and coordinates for point A of ($Y_1 = 6$, $Y_2 = 11$) and for point B of ($Y_1 = 9$, $Y_2 = 13$) the

(b)

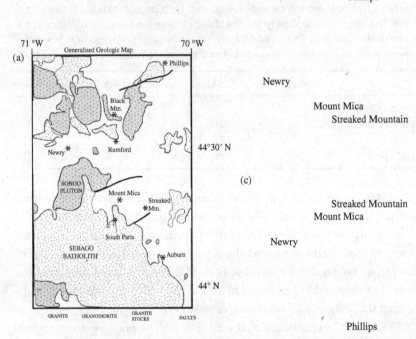

Phillips

Newry

Mount Mica
Streaked Mountain

(c)

Streaked Mountain
Mount Mica

Newry

Phillips

Figure 20.11 (a) The true geologic map and (b) and (c) two equally plausible multidimensional scaling plots showing the relationships between four pegmatites in terms of their distances apart. Both maps correctly show all the dissimilarities among the four pegmatites and are therefore equally applicable, even though only one (in this case (b)) corresponds to the actual position of these pegmatites on the geologic map.

distance between them will be the hypotenuse of a triangle which has sides 3 units long (9 − 6) on axis Y_1, by 2 units (13 − 11) high on axis Y_2. Therefore the length of the hypotenuse is the square root of (9 + 4) which is 3.61 units.

So the general formula for the Euclidian distance between any points whose coordinates are known in p dimensions is:

$$d_e = \sqrt{\sum_{i=1}^{p} (Y_{iA} - Y_{iB})^2} \qquad (20.1)$$

where $Y_1, Y_2, Y_3 \ldots Y_p$ are the number of dimensions. For example, for only two dimensions Y_1 and Y_2 are the X and Y axes of a typical two-dimensional

Table 20.6 Calculation of the Euclidian distance between two samples A and B, on which three variables have been measured. The samples can be positioned in a three-dimensional space in relation to their values for each variable. The Euclidian distance is the square root of the sum of the squared differences between samples for each of the three variables (in this case, three trace elements).

Variable	Sample A	Sample B	$(Y_A - Y_B)$	$(Y_A - Y_B)^2$
Sm (axis Y_1)	24	12	12	144
Eu (axis Y_2)	33	31	2	4
Gd (axis Y_3)	121	95	26	676
$\sum_{i=1}^{p} (Y_{iA} - Y_{iB})^2$				824
Single univariate value for the Euclidian distance between samples A and B				28.71

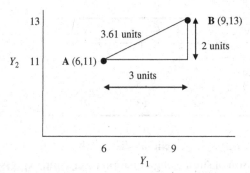

Figure 20.12 The Euclidian distance between two points, A and B, plotted in two dimensions.

scatter plot. The 'Y' terminology is used because the number of dimensions (and therefore the number of axes) can be two or more.

Equation (20.1) gives a **single value** for the dissimilarity between the two points, just like the distances between pegmatites previously described. If the values for each variable measured on each point are identical, the Euclidian distance (and dissimilarity) will be zero. Table 20.6 gives an example for three variables.

20.14 An example

The data in Table 20.7 are for six different trace elements measured on acid-sulfate waters extracted from four geothermal wells (A–D). By inspection of these raw data, it is hard to see which wells are most similar and which are

Table 20.7 Raw data for the concentrations of six different trace elements (in parts per million) at four geothermal wells.

Element	Site A	Site B	Site C	Site D
Ce	12	16	22	14
Nd	43	54	6	39
Eu	32	34	54	28
Tb	61	23	32	71
Ho	2	7	10	8
Tm	31	65	4	29

Table 20.8 The matrix of results for the Euclidian distances between all six possible paired combinations of sites shown in Table 20.7.

	Site A	Site B	Site C	Site D
Site A	0			
Site B	52.6	0		
Site C	59.9	80.9	0	
Site D	13.3	62.2	63.1	0

most dissimilar. In Table 20.8, the Euclidian distance has been calculated for each pairwise comparison of sites, using Equation (20.1), and expressed in a matrix.

The calculated matrix of dissimilarities can be used to position the sites in only two-dimensional space, as has been done in Figure 20.13. The process becomes difficult to do by hand as soon as you have more than three objects, but statistical packages are available to do this. Some of these simply start by placing the sampling units entirely at random in two dimensions. (At this stage the distances among them are extremely unlikely to correspond to the actual Euclidian distances.) Next, all of the sampling units are moved slightly at random. If this improves the correspondence between the positions of the sites within the two-dimensional space and their known Euclidian distances apart, then the change is retained. If it does not improve the fit, then the change is discarded and another change chosen at random.

This is done **iteratively,** which means it is repeated many (thousands or tens of thousands) times, and will result in a gradually improving map of the relationships among the sites. Eventually the fit cannot be improved any

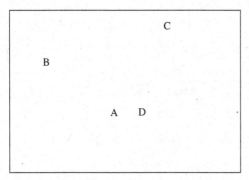

Figure 20.13 Example of the arrangement of the four sites (for which data are given in Table 20.7) in two dimensions on the basis of the Euclidian distance between each pair of sites.

further and the process stops. At this stage, there will be a final map showing the best relationship among the sites.

Importantly, there may be several possible final maps, so most MDS programs repeat the process several times to establish the most common solution.

20.15 Stress

Ideally, the display of dissimilarities will be two dimensional because this is easiest to interpret. Sometimes, however, the sites will not fit well into a flat two-dimensional plane, which will have to be rippled in certain places to place a site (or sites) so that they are the appropriate Euclidian distances from all the others. This lack of conformity to a two-dimensional display is called **stress** and will give a misleading picture of the relationships among sites. For example, a site forced up on a ripple will seem closer to two neighboring sites than it really is when the relationships are viewed as a two-dimensional display (Figure 20.14).

Stress can be reduced by increasing the number of dimensions and there will be no stress at all when the number of dimensions is equal to the number of original variables, but that is unlikely to be useful to you because a multidimensional display is usually impossibly complex to interpret. Hopefully you will get a display with little stress, in only two or three dimensions. Statistical packages that do MDS usually give a value for stress: as a general guide, it should be less than 0.2 and ideally less than 0.1.

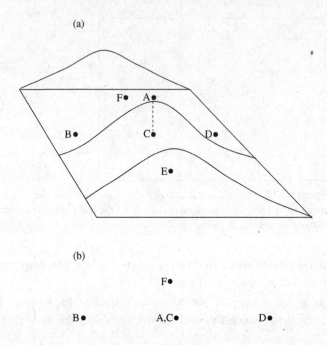

Figure 20.14 Sometimes sites will not fit into a two-dimensional plane. (a) Sites B, C, D and F are on the flat "floor" of the figure. Site A can only be accommodated accurately in relation to all others by positioning it in space above (or below) C. (b) Seen from above, as a two-dimensional map, A is misleadingly close to C.

20.16 Summary and cautions on the use of multidimensional scaling

Multidimensional scaling is a way of displaying sampling units, for which multivariate data are available, in a reduced number of dimensions. The distance between the sampling units is an indication of their dissimilarity. Unlike PCA, it does not identify which variables contribute to the positions of the sampling units.

Many different dissimilarity indices have been developed. For continuous data, where there are few values of zero, Euclidian distance is appropriate. MDS is frequently used by biologists and environmental scientists to analyze data for the numbers of several different species. Often these data sets

Table 20.9 (copied from Table 20.8). The matrix of results for the Euclidian distances between all six possible paired combinations of sites shown in Table 20.7.

	Site A	Site B	Site C	Site D
Site A	0			
Site B	52.6	0		
Site C	59.9	80.9	0	
Site D	13.3	62.2	63.1	0

include large numbers of zero values, so dissimilarity indices (e.g. the Bray–Curtis coefficient) which are not biased by the inclusion of zeros have been developed for these and should be used for **any** data set that contains a large proportion of zeros.

Although MDS is a simple technique for displaying sampling units in as few as two dimensions, the amount of stress (Section 20.15) required to do this needs to be considered, because the two-dimensional projection is likely to be misleading when stress is high.

20.17 Q-mode analyses: cluster analysis

Cluster analysis is a method for classifying sampling units into groups (called **clusters**) where those within a particular cluster are more similar to each other than they are to sampling units in other clusters. This is much simpler than it sounds. For example, the a posteriori Tukey test (Chapter 11) for assigning several means to groups, based upon the criterion of no significant difference among means within each group, is a simple univariate clustering method.

The following explanation of cluster analysis relies on an understanding of how univariate data for the dissimilarity between sampling units are derived from multivariate data, which was explained in Sections 20.13 and 20.14.

Just like MDS, cluster analysis uses a matrix of univariate dissimilarities between pairs of sampling units. For example, the data for the concentrations of six metals at four sites in Table 20.7 were used to construct the matrix in Table 20.8, which has been copied to Table 20.9. It gives the Euclidian distance between all possible pairs of four sites.

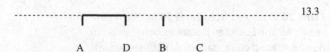

Figure 20.15 Fusion of the two most similar sites (A and D) to give three clusters on the basis of a maximum of 13.3 units of dissimilarity within clusters.

There are several types of cluster analysis, but one most commonly used is **hierarchical clustering** which can be used to construct a **dendogram** – a tree-like diagram – showing clusters based on **the amount of similarity within a cluster.** Here is an example.

First, to start with, the four sites in Table 20.9 can be considered as being in four separate groups or clusters, because none of the dissimilarities between any of them are zero.

Second, the dissimilarities in Table 20.9 are examined to find the two sites that are most similar. These are sites A and D, with a dissimilarity between them of only 13.3 units. These two sites are assigned to form the first cluster, with an internal dissimilarity of only 13.3 units (Figure 20.15).

At this stage the sites have been assigned to three clusters – one with A&D, plus B and C. The cluster of sites A and D, symbolized as (A&D), is then **considered as a single sampling unit** and the matrix of dissimilarities recalculated. This will not affect the dissimilarity between sites B and C, but the dissimilarity between site B and the "new" sampling unit of cluster (A&D), as well as that between site C and cluster (A&D) will change.

There are several methods for calculating dissimilarity after sites have started to be assigned to clusters. The **group average linkage method** simply takes the average of the dissimilarity measures between an outside sampling unit (e.g. site B) and those within the cluster (e.g. (A&D)). Therefore, using the initial dissimilarities given in Table 20.7, the new dissimilarity between site B and the cluster (A&D) is the **average** of the dissimilarity between B and A, and between B and D. This is $(52.6 + 62.2) / 2 = 57.4$. In the same way, the new dissimilarity between site C and cluster (A&D) is the average of the dissimilarity between C and A, and between C and D. This is $(59.9 + 63.1)/2 = 61.5$. (Note that the dissimilarity between B and C remains the same at 80.9.)

These dissimilarities will give the reduced matrix in Table 20.10. By inspection the two most similar sampling units are now cluster (A&D) and site B (because the dissimilarity is the lowest at 57.4). Therefore, these

Table 20.10 The reduced matrix of results for the Euclidian distances between the clusters shown in Figure 20.16.

	Sites (A&D)	Site B	Site C
Sites (A&D)	0		
Site B	57.4	0	
Site C	61.5	80.9	0

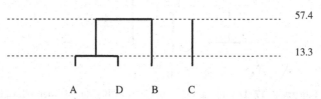

          57.4

          13.3

  A   D   B   C

Figure 20.16 Fusion of the first cluster (A&D) and the next most similar (site B) to give two clusters, (A&D&B) and site C, on the basis of a maximum of 57.4 units of dissimilarity within clusters.

two are now assigned to the same cluster, with an internal dissimilarity of 57.4. This gives two clusters: (A&B&D) and site C (Figure 20.16).

Next, the matrix of dissimilarities is reduced to the one in Table 20.11. Here, because there are only two sampling units left to compare, the only dissimilarity necessary to calculate is between (A&B&D) and site C. Here too, the dissimilarities in the very first matrix (Table 20.9) are averaged. The calculation is slightly more complex because you need to take the average of three dissimilarities A–C, B–C and D–C. This is $(59.9 + 80.9 + 63.1)/3 = 68.0$.

These values for increasing dissimilarity can be used to construct the final dendogram showing the sites grouped into fewer and few clusters (Figure 20.17). The dendogram shows a three-cluster solution at 13.3 internal dissimilarity, a two-category solution at 57.4 internal dissimilarity, and a single-category solution at 68.0 internal dissimilarity. This result is consistent with the results of the MDS analysis of the same data in Figure 20.13, which is not surprising.

The advantage of a cluster analysis is that it gives you **a quantitative way of assigning sampling units to groups**. For example, from the dendogram in Figure 20.17 you could suggest that A&D are "in the same group" which is different from group B and group C. Importantly, however, the groupings produced by a cluster analysis are unlikely to correspond to "true" categorical attributes (such as black or white sand grains) given as examples of nominal

Table 20.11 The matrix of results for the Euclidian distance between the only possible pair (A&D&B) and site C.

	Sites A&D&B	Site C
Sites A&D&B	0	
Site C	68.0	0

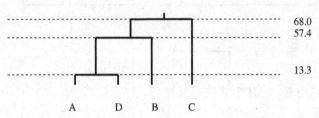

Figure 20.17 Dendogram showing sites A, B, C and D hierarchically arranged in fewer clusters as the amount of dissimilarity allowed within clusters increases. At 13.3 units of dissimilarity there are three clusters: (A&D), B and C. These reduce to two clusters: (A&D&B) and C at 57.4 units. Fusion into only one cluster occurs at 68.0 units of maximum dissimilarity within a cluster.

scale data in Chapter 6. Instead, the categories are based on decisions made about dissimilarity (or its converse, similarity) which is a **continuous and ratio scale variable**. This is an important point. Cluster analysis was primarily developed for taxonomists – biologists who describe and define animal and plant species – as a way of helping them make a decision as to whether individuals should be categorized as the same or different species. Here too, however, even though the analysis can be used to define clusters it does not mean these have identified real discontinuities or discrete categories.

In geological applications, cluster analysis has become a commonly used technique to distinguish and help characterize groups on the basis of many types of geological data: major, minor, and isotope geochemistries, sediment particle sizes, drainage basin morphologies, or fossil contents. Cluster analysis has even been used to group different boulder morphologies and differentiate geochemical units on Mars. In these applications, results of cluster analysis are often highly dependent on normalization of the data for the respective variables, particularly because geological data may lack normal or log-normal distributions and be strongly skewed or have multiple modes.

20.18 Which multivariate analysis should you use?

The three analyses described in this chapter are all ways of summarizing and simplifying a multivariate data set so that relationships among sampling units can be more easily visualized, but they have different applications.

Principal components analysis is useful for data sets where there are few zero values and you need to know which variables contribute most to differences among sampling units.

Multidimensional scaling can be used with data sets that contain a lot of zero values. Most MDS programs do not give an indication of which original variables contribute to differences among sampling units.

Cluster analysis assigns objects to groups, based on dissimilarity or similarity, which may help you categorize the sampling units.

This chapter has only described three commonly used multivariate methods. This is deliberate. First, many earth scientists may never use multivariate analyses themselves, but will need a conceptual grasp of how they actually work, so they can evaluate reports that include summary statistics and conclusions from multivariate data. Second, more powerful methods of analyzing multivariate data are being developed, but most of these are derivations of these three "core" methods.

20.19 Questions

(1) Discuss the statement "If there are no correlations within a multivariate data set then principal components analysis really is not very much use at all."

(2) An earth scientist carried out a principal components analysis and obtained the following eigenvalues for components 1 to 5. Which components would you use for a graphical display of the data? Why?

Principal component	Eigenvalue	Percentage variation
1	3.54	54
2	2.82	23
3	2.64	22
4	0.89	6
5	0.42	5

(3) What is "stress" in the context of a two-dimensional summary of the results from a multidimensional scaling analysis?

(4) Why are the "groups" produced by cluster analysis often not equivalent to true cases of categorical data (such as black versus white sand grains)?

21 | Introductory concepts of sequence analysis

21.1 Introduction

Geoscientists often have to interpret data that are in the form of a sequence – an ordered series of observations – that has been measured over time or space. For example, on a **temporal scale** you might have data for sea level that has been repeatedly sampled at the same location over several months, years, or decades and need to know if the mean has changed, whether there is a consistent trend, or even repetition of the same pattern. The analysis of temporal scale data is often called **time series analysis**. On a **spatial scale**, sequential data might be obtained as a core from a bore hole drilled down through a sedimentary sequence or a stack of lava flows. Although such sequences are spatial, they could also be thought of as temporal because deeper rocks are likely to be older, but depth and age are unlikely to be equivalent because the thickness deposited may vary among years. Nevertheless, the same statistical methods can often be used for both temporal and spatial sequences.

Data for a sequence can be measured on a ratio, interval scale or ordinal scale (e.g. the conductivity of well water over several months) or a nominal scale (e.g. chemical or porosity changes with depth in a stratigraphic sequence).

Analysis of a sequence might detect a trend (or a lack of it), or reveal features that may lead to hypotheses about temporal or spatial processes. Patterns of occurrence within a sequence may also be used as predictors of conditions of interest. For example, deposits of some minerals (e.g. uranium) are accompanied by very characteristic modifications of the geochemistry of surrounding rocks. The resultant alteration haloes are especially characteristic of minerals formed by migrating uranium-rich fluids. It would be very useful to know that areas of uranium mineralization had a 40%

probability of occurring below a particular type of rock showing an alteration halo (e.g. bleaching of initially hematite-rich sandstone).

All of the techniques for sequence analysis described here use statistical methods explained earlier in this book. We will assume an understanding of correlation (Chapter 15), regression (Chapter 16) and contingency tables (Chapter 18) to introduce the essential concepts, terminology and techniques of sequence analysis and interpretation.

21.2   Sequences of ratio, interval or ordinal scale data

A sequence of ratio, interval or ordinal scale data measured temporally or spatially is a bivariate data set with a **measured variable** (e.g. sea level) and a **sequence variable** (e.g. time or distance) giving position within the sequence.

Several things may affect the measured variable. First, there is likely to be a random component (the "error" discussed in Chapters 10 and 16). Second, there may be a longer-term upward or downward trend. Third, there may be a regular repetitive pattern such as the annual summer/winter fluctuation in temperature, or a longer-term repetition (e.g. climate change) that is not annual or seasonal. Fourth, part(s) of the sequence may be consistently higher or lower than the mean. Finally, the value of the measured variable may be somewhat dependent on the value(s) in previous parts of the same sequence. A sequence analysis is used in an attempt to explain as much of this variation as possible in order to characterize a sequence, test for significant variation over time and perhaps even make some very cautious predictions.

21.3   Preliminary inspection by graphing

As a first step, it is very helpful to graph the measured variable on the Y axis and the sequence variable (e.g. time) on the X axis. For example, Figure 21.1 gives the strength of the magnetic field of the Earth during the past 100 years. Many scientists interpret this decrease in the dipole moment to be a precursor to a reversal of the Earth's magnetic poles.

By inspection, the decrease in field strength is approximately linear. Both variables have been measured on a ratio scale, so the first (and simplest) model applied to the data could be a linear regression with field strength (Y) as the dependent variable and time (X) as the independent one

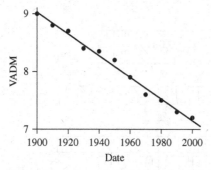

Figure 21.1 Strength of the Earth's magnetic field expressed as the virtual axis dipole moment (VADM as 10^{22} Am^2) during the past century.

(Chapter 16). If the regression line appears to be a good fit to the data and the assumptions of regression are met, it may be all you need to describe the sequence and test for a significant change in the measured variable over time.

Most sequences are more complex than the one in Figure 21.1. Often the relationship between the measured variable and the sequence variable is not linear, and there may be **similarity or dissimilarity** between different parts of the sequence.

21.4   Detection of within-sequence similarity and dissimilarity

As a second exploratory step to help establish the features of a sequence, it is often examined for **within-sequence similarity and dissimilarity.** As an example, consider an ice core from a glacier, where the percentage of impurities has been measured at regular intervals down the length of the core. Any repetition of the same or similar values, or pattern (e.g. a regular cyclic change) along the length of the core may help understand the processes responsible for changes within a sequence and can even be used to tentatively predict what might happen in the future.

One way of detecting repetition is to copy the data from the core, thus giving two identical sequences. If these two sequences are laid parallel to each other and side by side, with the beginning of the "top" sequence aligned with the beginning of the "bottom" one, then each of the adjacent values in the two sequences will be the same (Figure 21.2(a)).

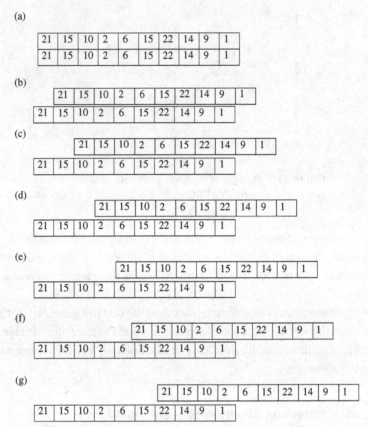

Figure 21.2 Examination of a sequence, running from left to right with the most recently recorded value on the right, for internal similarity and dissimilarity. (a) The sequence of data on percent impurity is copied and placed alongside itself to give two identical sequences. (b) The top sequence is shifted by one interval to the right, thereby putting every value in the lower sequence adjacent to that for the previous interval in the top one, and the overlapping sections compared. (c)–(g). The process described in (b) is repeated. For the shift shown at (g), the two sets of four cells in the overlapping section have similar values, thus indicating a pattern of similarity between different parts of the sequence. Note also that for the shift in (d), high values in one sequence are aligned with low values in the other, indicating sections where the pattern in one is the opposite (and therefore markedly dissimilar) to the other.

Next, the top sequence is successively shifted to the right, by one observation at a time. After each shift the overlapping parts of the two cores are compared to each other to see if they are similar or dissimilar (Figure 21.2(b–g)). As the two cores are progressively moved past each other, the most recent parts of the bottom core will occur adjacent to older and older parts of the top one, so if a pattern occurs within a sequence then the similar or dissimilar sections will, at some stage, lie side by side (Figure 21.2(g)).

This method is straightforward, but **an essential assumption is that samples have been taken at regular intervals throughout the sequence** (e.g. usually an equal length increment in geological settings). If the intervals are unequal, then it may be possible to obtain a regular sequence by excluding some data.

It would be very time consuming to visually inspect the two sequences every time they were shifted. Furthermore, you need some way of **deciding whether any similarity or dissimilarity is significant** or whether it might only be occurring by chance within a sequence of random numbers. This can be done by using **autocorrelation** (which is sometimes called **serial correlation**) to test for a relationship, without assuming dependence or causality. As described above, a sequence is copied to give two identical ones which are then placed side by side (Figure 21.2(a)). The values adjacent to each other will be the same, so at this stage the correlation (Chapter 15) between the variables "sequence 1" and "sequence 2" will always be 1.0.

Next, sequence 1 is shifted only one interval to the right (Figure 21.2(b)). This shift is called a **lag interval of one** (or just a **lag of one**), and it places every value within sequence 2 adjacent to the value recorded at the previous interval in sequence 1. The correlation is recalculated. The process is repeated several times: the sequence is shifted another interval in the same direction (therefore giving lag intervals of two, three, four etc.) and the correlation recalculated each time (Figure 21.2(c)–(g)). The number of lags that can be used will be limited by the length of a finite sequence, because every successive shift will reduce the length of the overlapping section by one.

If there is marked **similarity** within the sequence then the correlation at some lag intervals will be strongly positive (e.g. Figure 21.2(g)).

If there is **no marked similarity or dissimilarity** and only random variation, the correlation will show some variation but have a mean of zero.

If the **pattern at a particular lag in one sequence is the opposite of the other and therefore markedly dissimilar,** the correlation will be strongly negative (e.g. Figure 21.2(d)).

To obtain Pearson's correlation coefficient for a set of bivariate data (Chapter 15), the means of each variable are separately calculated and used to convert the two sets of data to their Z scores using the following formulae.

For a population:

$$Z = \frac{X_i - \mu}{\sigma} \qquad \text{(21.1 copied from 15.1)}$$

and for a sample:

$$Z = \frac{X_i - \bar{X}}{s} \qquad \text{(21.2 copied from 15.2)}$$

Importantly, a sequence assessed for autocorrelation is usually treated as a **population** because it contains all of the data for that sequence. Therefore, when calculating Z scores **the mean and variance of the entire sequence are used**, not just the sample means and variances for the overlapping sections.

Using Z scores, the Pearson correlation coefficient for a population is:

$$r = \frac{\sum_{i=1}^{N}(Z_{xi} \times Z_{yi})}{N} \qquad (21.3)$$

For autocorrelation (that compares the measured variable to itself) the use of Z_x is inappropriate and Z_y is used instead:

$$r = \frac{\sum_{i=1}^{N}(Z_{yi} \times Z_{yi})}{N} \qquad (21.4)$$

Equation (21.4) gives the autocorrelation for a lag of zero, but this will always be 1.00. As the lag interval increases the number of overlapping values will decrease so the actual number of values being correlated will be fewer (Figure 21.2).

To calculate the autocorrelation between different lags of the same sequence, two modifications to Equation (21.4) are needed.

First, to specify the actual parts of the sequences being compared, the numerator of Equation (21.4) is changed to that shown in Equation (21.5), where k is the lag number. This may look complex, but working through the equation using an example will help. For a lag $k = 10$ and $i = 1$, then Z_{yi} (which is $Z_{y(1)}$ and the first value in the sequence), will be paired with $Z_{y(i+k)}$ (which is $Z_{y(11)}$ and the 11th value in the sequence). For the same lag of 10, when $i = 2$ the numerator will pair Z_{yi} (which is $Z_{y(2)}$) with $Z_{y(i+k)}$ (which is

$Z_{y(12)}$) etc. This ensures that the appropriate Z scores are multiplied together.

Second, because the number of values being correlated is the total within the sequence minus the lag number (e.g. at lag 0, for a sequence of length 50, all 50 values will be used, but a lag of 5 will use only 45), the denominator of the equation becomes $N - k$ where k is the lag number. Note also that the value above the symbol Σ is also $N - k$ which restricts the Z scores being used to those for the overlapping sections of the two cores (Figure 21.2).

$$r = \frac{\sum_{i=1}^{N-k} Z_{yi} \times Z_{y(i+k)}}{N - k} \tag{21.5}$$

Once the values for the correlation coefficient at each lag interval have been calculated, they are plotted as a line graph **with r on the Y axis and the lag number on the X axis.** This graph is called a **correlogram** and several examples are given in Figure 21.3. The correlation coefficient at lag zero will always have an r of 1.0, which is why correlograms produced by statistical packages often only plot lag intervals of one and more.

21.4.1 Interpreting the correlogram

The shape of the relationship between the Pearson correlation coefficient r plotted against lag is a very good indication of the characteristics of the sequence.

A sequence that shows **no overall trend and only random variation with no marked internal similarity or dissimilarity** will have a value of r that starts at 1.0 for lag zero, but will very rapidly decrease and has an expected average correlation of $r = 0.0$ at all higher lags (Figure 21.3(a)). This is an example of a **stationary sequence** because the original sequence variable shows no overall upward or downward trend.

If the value of the variable has **some dependence on the value in the previous interval or intervals** (i.e. the value for Y_t is related to that for Y_{t-1} or even Y_{t-2} and Y_{t-3}) then r will show strong positive or strong negative autocorrelation at low lags but an average of zero for higher ones (Figure 21.3(b)).

A **trend over time, whether it is decreasing or increasing,** will give a value of r that starts at 1.0 but then slowly decreases to a marked negative

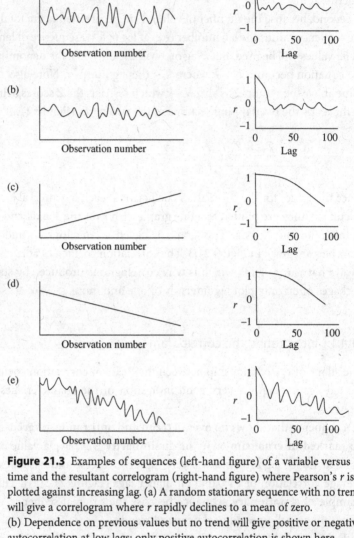

Figure 21.3 Examples of sequences (left-hand figure) of a variable versus time and the resultant correlogram (right-hand figure) where Pearson's r is plotted against increasing lag. (a) A random stationary sequence with no trend will give a correlogram where r rapidly declines to a mean of zero.
(b) Dependence on previous values but no trend will give positive or negative autocorrelation at low lags: only positive autocorrelation is shown here.
(c), (d) An increasing or decreasing linear trend will show marked positive autocorrelation at low lags, but marked negative autocorrelation at high lags, the latter because as lag increases the similarity between the Z scores in the overlapping sections decreases to the point where they are markedly dissimilar. (e) Decreasing trend, with random variation superimposed.
(f) A regular cyclic component will give a regular pattern in the correlogram.
(g) When there is a trend plus within-sequence repetition, the correlogram will show a gradual decrease as well as fluctuations caused by the repetition.

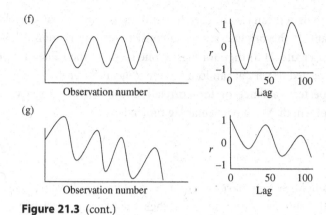

(f)

Observation number

Lag

(g)

Observation number

Lag

Figure 21.3 (cont.)

correlation as lag increases (Figure 21.3(c) and (d)). These are **non-stationary** sequences because the original variable shows an overall trend.

If there is **a consistent positive or negative trend, plus random variation** (Figure 21.3(e)) then r will fluctuate but will be markedly positive at low lags, steadily decrease as lag increases and eventually become markedly negative. Here too, the sequence is non-stationary.

If there is **no overall trend but regular repetition of similar or dissimilar sections within a sequence**, then the correlogram will show autocorrelation at regular lag intervals (Figure 21.3(f)). In this example, even though there is fluctuation, there is no overall long-term positive or negative trend, so the series is stationary.

Finally, if there is **a long-term positive or negative trend, plus repetition within the sequence** (Figure 21.3(g)), then the correlogram will show marked positive autocorrelation at low lag intervals and markedly negative autocorrelation at high ones, but will also fluctuate because of the repetition. This is a good example of how two sources of variation can affect the value of r.

In summary, the amount of autocorrelation will be affected by (a) random variation, (b) the strength of any long-term trend in non-stationary sequences and (c) whether there is similarity among different parts of a sequence. Therefore, when both (b) and (c) are present the values of r in some parts of the correlogram can be misleading (e.g. Figure 21.3(g)) and it is necessary to remove the long-term trend in order to assess the extent of repetition. This is discussed later in the chapter.

The correlogram can also be used to test if the amount of autocorrelation is significant. If the original sequence consists of only random variation (case (a) in Figure 21.3) then for lags of one or more the value of r would only be expected to vary at random around a mean of zero.

The expected **variance** of the correlation coefficient r for a random sequence of length N at a particular lag time k is:

$$\sigma_r^2 = \frac{1}{(N - k + 3)} \tag{21.6}$$

For example, if you have a sequence containing 40 values and you calculate the autocorrelation at lag 4, then the expected variance at that lag is: $1/(40 - 4 + 3)$, which is $\sigma_r^2 = 0.0256$. From Equation (21.6) it is clear that the variance is affected by the sequence length (for a short sequence the expected variance will be large, but will decrease as N increases), and the amount of lag (as k increases the variance will increase).

The expected standard deviation of r is just the square root of Equation (21.6). For a population, 95% of the values of the correlation coefficient are expected to fall within 1.96 standard deviations of $r = 0$:

$$0 \pm 1.96 \times \sqrt{\frac{1}{(N - k + 3)}} \tag{21.7}$$

so if the value of r is outside this range it shows significant autocorrelation at $P < 0.05$.

The 95% confidence limits can be drawn on the correlogram as two curved lines, with significant autocorrelation occurring whenever r is outside this range. **Importantly, any test of the significance of r will only give a realistic result when the sequence is relatively long (e.g. at least 40–50 observations) and the number of lags for which r is calculated are relatively few.** This is because the length of the overlapping sections will get smaller and smaller as lag is increased, so the correlation will be between shorter and shorter parts of the sequence, as shown in Figure 21.2. Therefore, it is **recommended that values of r are only calculated for lags up to one quarter of the full sequence length.** Despite this, statistical packages often give autocorrelations for every possible lag of even short sequences, so you need to be extremely cautious about the reliability of statistics for lag numbers more than about one quarter of any sequence length.

The formula for the autocorrelation given here is probably the easiest to understand but there are several variations, including ones that treat the sequence as a sample and not a population. All will give similar results as long as the test is limited to the first quarter of a relatively long sequence. Most statistical packages will give a graph of r and its 95% confidence limits, and there are examples in the following section.

Some statistical packages also include a table showing **the Box–Ljung statistic** (that some texts and web pages call the Ljung–Box statistic), which indicates the extent of autocorrelation for the combined set of lags up to and including the one for which the Box–Ljung statistic is given. For example, the Box–Ljung statistic at lag 10 gives the extent of autocorrelation within lags 1–10 inclusive, and you still need to examine the correlogram to identify which ones are significant. The formula for the Box–Ljung statistic is:

$$Q = N(N+2) \sum_{k=1}^{h} \frac{r_k^2}{N-h} \qquad (21.8)$$

where N is the number of values in the original sequence, k is the lag number, h is the maximum lag number for the range being tested and r_k is the autocorrelation at each lag. The size of Q is affected by the cumulative amount of autocorrelation within the sequence up to the point at which it is calculated.

21.5   Cross-correlation

Cross-correlation is very similar to autocorrelation, but is used to compare **two different sequences**, which may even be for different variables. Therefore, the two series are unlikely to show perfect correlation at lag 0. For example, you might want to compare data for the flow discharge of water in a stream with the water use patterns at a nearby golf course for the same (or a longer) time period to see if there is any relationship (and if so, what the lag is) between these, in order to know how long it takes for irrigation to affect discharge.

For cross-correlation, the method for obtaining the correlation coefficient at different lags is similar to the one described above, but because two different sequences are being compared and the comparison is usually restricted to parts of each sequence, the overlapping sections are treated as samples.

21.6   Regression analysis

A sequence of ratio, interval or ordinal scale data can often be analyzed by regression, provided the assumptions of this procedure are met (Chapter 16).

First, the characteristics of the sequence are determined by exploratory testing, including autocorrelation, as described above. A regression model is chosen, fitted to the sequence and assessed to see if it is appropriate. If necessary, the model is refined. The assessment and refinement steps may have to be repeated several times to develop a model to the stage where it is a good description. Finally, the model is used to draw conclusions about the sequence. These steps are summarized in Figure 21.4.

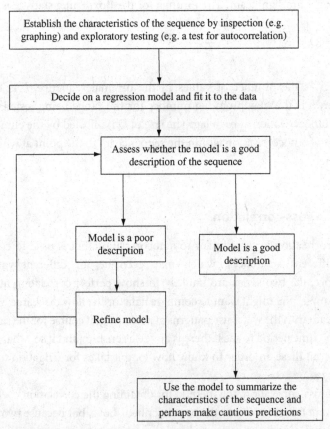

Figure 21.4 The general steps for using regression to analyze a univariate sequence.

The coefficient of determination, r^2, shows how much of the variation is explained by the regression. A plot of the **residuals** will help indicate if the regression model is an appropriate analysis for the data (Chapter 16).

The residuals $(y_i - \hat{y})$ have another use in sequence analysis. By plotting them against the independent variable X, the regression line of best fit $(\hat{y})$ is converted to a horizontal line where all values of $\hat{y}$ are equal to zero, with the residuals dispersed about it. This effectively **removes the variation explained by the line of best fit from the sequence**, and has two advantages. First, if the regression line is a good fit to the data, the plot of the residuals will now be independent of any long-term trend. This is one way of **detrending** a sequence, and the detrended data can be used to investigate autocorrelation caused by **repetition within the sequence** without the confounding effect of any general trend upon the value of r.

Second, if the residuals are not evenly distributed about zero it suggests there is still variation present that is not accounted for by the line of best fit. The pattern of the residuals about the line may indicate the characteristics of this variation, from which you could make a choice of additional terms to incorporate into the regression equation in an attempt to improve the fit of the model.

A further step often used in sequence analysis is to draw a **correlogram of the residuals**. If the regression is a good description of the data, the values of r at lags of one or more in this correlogram should only show random fluctuation around a mean of zero. Significant values of r will indicate any remaining autocorrelation.

Choosing an appropriate model requires a good understanding of complex regression. Statistical packages can do extremely complex autoregression analyses, but these models have quite stringent assumptions and are often misapplied and misinterpreted. Therefore, if the sequence appears complex it is important to seek expert advice. Here we give straightforward examples of the use of some regression models.

21.7   Simple linear regression

For a sequence that shows an apparently linear trend over time, as in Figure 21.1, the correlogram should be similar to Figure 21.3 (c) or (d), but you might not even draw one for such an obvious relationship. The sequence could be analyzed using simple linear regression:

$$Y_i = a + bX_i \qquad\qquad (21.9 \text{ copied from } 16.1)$$

where Y is the measured variable, a is the intercept, b is the slope and X is the sequence variable (e.g. time). A test of the slope of the line will show if there is a significant change in Y over the sequence. You then need to check that the data satisfy the assumptions of regression, including whether the residuals, $(y_i - \hat{y}_i)$, show a relatively even spread around zero when plotted against X (Chapter 16).

Here is an example using turbidity – the opacity of a fluid caused by the presence of small particles in suspension – which can be estimated with a nephelometer that measures the amount of scattering when a beam of light is shone through fluid. The units of turbidity are called Nephelometric Turbidity Units (NTU). Figure 21.5(a) gives the turbidity

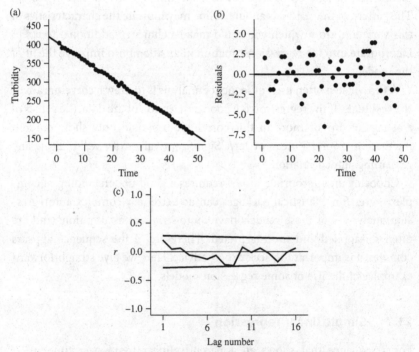

Figure 21.5 (a) Regression line, (b) unstandardized residuals and (c) correlogram of the unstandardized residuals, including 95% confidence limits, for the turbidity of water, measured in NTU, from well TGM006 in central Queensland at monthly intervals from January 2000 to December 2003.

of water in well TGM006 for four years after it was drilled in central Queensland in January 2000. By inspection, a straight line with a negative slope is likely to be a good description of the relationship (and this was so obvious that a correlogram was not drawn). The regression line $Y = 407 - 5.02X$ shown in Figure 21.5(a) has a highly significant negative slope ($F_{1,46} = 36984.8$, $P < 0.001$) and explains almost all of the variation, because $r^2 = 0.999$.

The unstandardized residuals plotted against time in Figure 21.5(b) show a fairly even distribution about zero. Finally, a correlogram of these residuals (Figure 21.5(c)) does not show any remaining autocorrelation and thus no significant additional variation. In summary, simple linear regression appears to be a good model of the data and shows a significant decrease over time. From this you could even cautiously forecast turbidity in future years, although your prediction may not be correct.

21.8   More complex regression

Often the relationship between the measured variable and the sequence variable is not linear. One way of modeling a more complex relationship is to expand the simple linear regression Equation (21.9) by adding additional constants and powers of X. These equations are called **polynomials** of increasing degrees.

The simple linear regression equation is a first-degree polynomial that gives a straight line relationship:

$$Y = a + b_1 X \qquad \text{(21.10 modified from 21.9)}$$

that can be expanded to a second-degree (quadratic) polynomial which gives a line with one change of direction, by adding a second constant that is multiplied by the square of X:

$$Y = a + b_1 X + b_2 X^2 \qquad (21.11)$$

and a third-degree (cubic) polynomial that gives a line with two changes of direction, by adding a third constant multiplied by the cube of X:

$$Y = a + b_1 X + b_2 X^2 + b_3 X^3 \qquad (21.12)$$

and a fourth-degree (quartic) polynomial that gives three changes in direction:

$$Y = a + b_1 X + b_2 X^2 + b_3 X^3 + b_4 X^4 \qquad (21.13)$$

and so on, for additional constants (b) and increasing powers of X.

As the number of constants and powers of X increases, the regression line will become a better and better fit to the data. Eventually, when the number of terms in the polynomial is one less than the number of data points within the sequence, the equation will run through all the points (and thus be a perfect fit), but this is unlikely to be useful for anything except a very short sequence because the equation will be extremely long and complex. Often, however, a good approximation of a long-term trend can be achieved by using only a second- or third-degree polynomial, which may also detrend the sequence. The regression can be tested for significance, easily visualized and used for interpolation and prediction.

To illustrate the successively better fit of increasingly complex regression models, Figure 21.6 shows turbidity of water from a second well (TGM013), where the relationship is clearly not linear. First, a linear model using Equation (21.10) does not show a significant relationship between turbidity and time

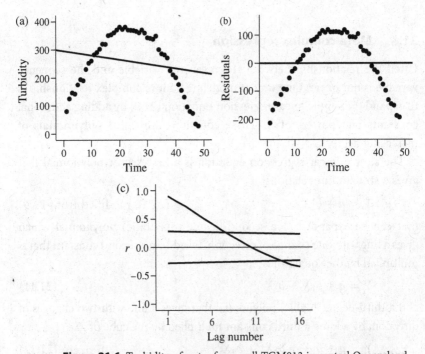

Figure 21.6 Turbidity of water from well TGM013 in central Queensland from early 2000 to late 2003. (a) A simple linear regression (heavy line) is clearly not a good fit to the points. (b) Unstandardized residuals. (c) Correlogram of the residuals for a linear model fitted to the data. The residuals are not evenly distributed and there is extreme autocorrelation.

($F_{1,46}$ = 2.008, NS) and the regression line (Y = 295.26 – 1.56 X) is an extremely poor fit (Figure 21.6(a)) with an r^2 of only 0.042. A plot of the unstandardized residuals (Figure 21.6(b)) confirms the use of linear regression is inappropriate because the data do not occur in a band around the horizontal line for zero residual variation. Nor has the regression detrended the data; the plot of residuals is similar to the original sequence with a pronounced long-term non-linear trend. Finally, the correlogram of the residuals in Figure 21.6(c) also shows highly significant correlation at low and high lags, which confirms that there is remaining variation not explained by the regression.

Second, a quadratic model (Equation (21.11): Y = 50.28 + 27.8 X – 0.60 X^2) fitted to the data is highly significant ($F_{2,45}$ = 2122.84, P < 0.001), and a far better fit than the linear model, with an r^2 of 0.989 (Figure 21.7(a)). A graph of the residuals (Figure 21.7(b)) shows that the regression appears to be a very good fit to the data, and the correlogram of the residuals (Figure 21.7(c)) confirms this, with no significant autocorrelation at any lag.

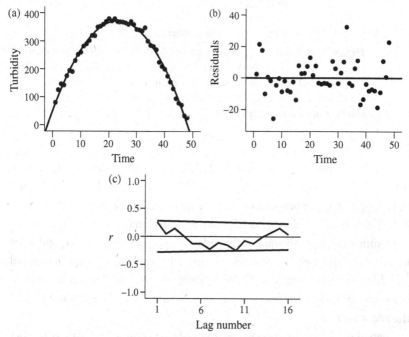

Figure 21.7 Turbidity of water from well TGM013 in central Queensland from early 2000 to late 2003. (a) A quadratic regression (heavy line) is a good fit to the points. (b) Unstandardized residuals. (c) Correlogram of the residuals for a quadratic model fitted to the data does not show significant autocorrelation.

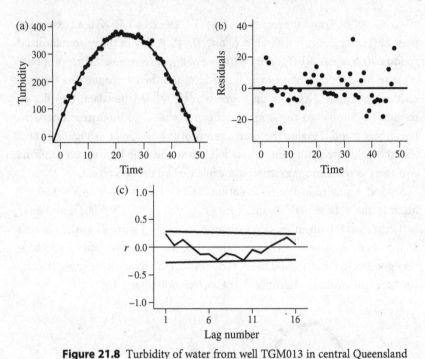

Figure 21.8 Turbidity of water from well TGM013 in central Queensland from early 2000 to late 2003. (a) A cubic regression (heavy line) is a good fit to the points, but appears little better than the quadratic in Figure 21.7. (b) Unstandardized residuals. (c) The correlogram of the residuals does not show significant autocorrelation.

Finally, a cubic model (Equation (21.12): $Y = 53.63 + 27.06\,X - 0.56\,X^2 - 0.001\,X^3$) is also significant ($F_{3,44} = 1392.67$, $P < 0.001$) with an r^2 of 0.990, indicating a slight improvement over the quadratic and a very good fit to the data. The residuals and their correlogram are consistent with this (Figure 21.8).

In summary, both the quadratic and cubic polynomials are very good fits to the data and can account for almost all the change in the measured variable over the sequence. There is a significant non-linear relationship between turbidity and time, with high turbidity during the middle part of the sequence.

Even though the regression fits the data, it cannot be used to predict turbidity in the future because the values will be negative and negative turbidity does not exist. We have deliberately chosen this example to illustrate the danger of predicting beyond the measured limits of a sequence, but the same caution applies even when the predicted values seem realistic.

From the value of r^2, the linear model is a very poor fit to the data. In contrast, the quadratic is a good fit and the cubic model is a slight improvement over the quadratic. There is no point in using a more complex higher-order polynomial if it does not give a significantly improved fit over a simpler one, and this can be tested for significance by a straightforward extension of the ANOVA used to assess the significance of a regression (Chapter 16). Most statistical packages give a table of results for the ANOVA that tests for departure from a line with zero slope (Chapter 16), which includes the sum of squares, degrees of freedom and mean squares for the regression. Table 21.1(a) gives these for the linear, quadratic and cubic models fitted to the data in Figures 21.6 to 21.8.

Each expansion of the polynomial is additive (e.g. Equations (21.10) to (21.13)) and so are their sums of squares. Therefore, the sum of squares for the improvement (if any) of the quadratic compared to the linear regression can be obtained by subtracting the sum of squares for the linear model from the sum of squares for the quadratic, giving the sum of squares for the difference (*SS difference*). The number of degrees of freedom for the difference (*df difference*) is also calculated by subtraction (Table 21.1(b)). The mean square for the difference is (*SS difference/df difference*), and the *F* statistic is calculated by dividing this quantity by the error of the higher polynomial (Table 21.1).

The same method is used to assess whether the cubic model is an improvement compared to the quadratic. In the example in Table 21.1 the additional variation explained by the quadratic over the linear model is highly significant, but the cubic model is not a significant improvement over the quadratic, so the latter is used to describe the relationship.

21.8.1 Polynomial modeling of a spatial sequence: hydrogen diffusion in a single crystal of pyroxene

There are many common geological phenomena where polynomial approximations are appropriate, particularly spatial data such as gravity models, porosity, and other fundamental rock properties that vary with depth, shoreline changes, as well as distortion and translation corrections in image analysis. For example, different types of diffusion processes are often described with polynomials. Figure 21.9 shows data from an experiment to measure hydrogen

Table 21.1 The amount of additional variation explained by progressive polynomial expansions can be assessed by subtracting the sum of squares for the lower polynomial from the next higher one to give the sum of squares for the difference (*SS difference*). The number of degrees of freedom for the difference (*df difference*) is also obtained by subtraction. The mean square for the difference is (*SS difference/df difference*), and the *F* statistic is calculated by dividing the resultant MS difference by the error of the higher polynomial. (a) ANOVA statistics for each of the three regression models. (b) ANOVA table for the relative importance of each additional expansion of the regression equation. In this example the additional variation explained by the quadratic model is a highly significant improvement over the linear one, but there is no significant improvement of the cubic model over the quadratic.

(a)

Model	Sum of squares	df	Mean square
Linear	(a) 5568.2	1	5568.2
Error	375 390.5	46	8160.7
Quadratic	(b) 530 847.9	2	265 423.9
Error	5653.1	45	125.6
Cubic	(c) 530 907.8	3	176 969.3
Error	5593.2	44	127.1

(b)

Model	Sum of squares for the difference	df	Mean square of difference	Error	df	F ratio
Quadratic minus	(b) 530 847.9	2		Quadratic		$F_{1,45} = 4182.2$
linear	(a) 5 568.2	1				$P < 0.001$
Difference	525 279.7	1	525 279.7	125.6	45	
Cubic minus	(c) 530 907.8	3		Cubic		$F_{1,44} = 0.47$
quadratic	(b) 530 847.9	2				NS
Difference	59.9	1	59.9	127.1	44	

diffusion in a single crystal of pyroxene. A water-rich crystal was cooked in a furnace so that hydrogen could diffuse out. After the experiment, the crystal was sliced open, and the concentration of hydrogen measured from edge to edge. The relationship between hydrogen concentration and location in the crystal is best described by a third-order polynomial (Figure 21.9).

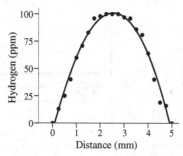

Figure 21.9 Hydrogen concentration (in ppm) measured from edge to edge across a single olivine crystal, after a dehydration experiment to characterize the diffusivity of hydrogen. A cubic regression (heavy line) is a good fit to the points. The data and figure have been simplified from Woods *et al.* (2000).

21.9   Simple autoregression

Often a correlogram will show significant autocorrelation at low lags. This may be because the series is non-stationary (e.g. Figure 21.3(d)), but for a stationary series, or one that has been detrended, low lag autocorrelation means there is a consistent **relationship between successive values**. For example, a sequence might show a significant dependence for the value of the measured variable at time (t) on time ($t-1$) (where t can be any point within the sequence). This dependence can be modeled by **autoregression** (meaning that the data set is regressed upon itself) and is of particular interest in fields such as finance and meteorology, where sequence analysis is used (with somewhat mixed success) in attempts to predict future values.

Here is an example. In some parts of the world annual rainfall has been found to be significantly related to annual rainfall during the previous year, or years (e.g. the value of the measured variable at time (t) is related to that at ($t-1$) or even ($t-2$) or ($t-3$)). Knowledge of this relationship, and its reliability, might help predict future rainfall: Figure 21.10(a) shows annual rainfall at the Neostrata 4 open cut coal mine in Western Australia from 1961–2008. During years when annual rainfall exceeds 350 mm the output of the mine has to be reduced because of the accumulation of storm water runoff in the floor of the cut, so some indication of the rain expected during the next year would help in planning extractive operations. Figure 21.10(a) shows fluctuation but no obvious longer-term trend, and

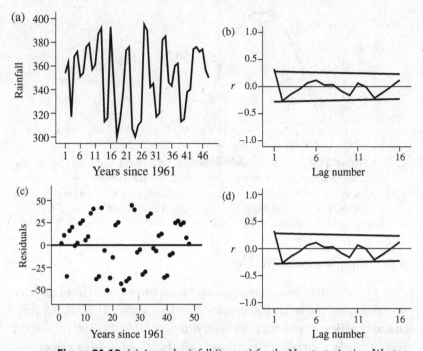

Figure 21.10 (a) Annual rainfall (in mm) for the Neostrata 4 mine, Western Australia, from 1961–2008 inclusive. (b) Correlogram, of original data. (c) Residuals from a linear regression. (d) Correlogram of the residuals. Note that (b) and (d) are extremely similar, thus showing that the linear regression accounts for little variation in the data.

therefore a stationary series, which was confirmed by the linear regression having a slope close to zero ($Y = 352.26 - 0.072\,X : F_{1,46} = 0.059$, NS). The regression is an extremely poor fit ($r^2 = 0.001$). A correlogram of the original data shows significant positive autocorrelation at lag 1 and an almost significant negative autocorrelation at lag 2 (Figure 21.10(b)). A scatter plot of the residuals against time is relatively evenly spread, but nevertheless shows a pattern where the values for successive years are often similar (Figure 21.10(c)). This was confirmed by the correlogram of the residuals, which is almost the same as the one for the original data (Figure 21.10(d)).

First, considering the autocorrelation at lag 1, a regression equation which only included autoregression on the previous year's rainfall was fitted to the data:

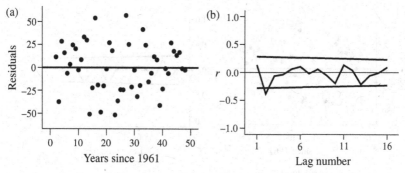

Figure 21.11 Autoregression analysis at $t-1$ for the rainfall data at Neostrata 4. (a) Residuals after autoregression at $t-1$. (b) Correlogram of the residuals. There is significant negative autocorrelation remaining at lag 2.

$$Y_t = a + B_1 Y_{t-1} \qquad (21.14)$$

which was significant $(Y_t = 237.08 + 0.323Y_{t-1}: F_{1,45} = 5.26, \ P < 0.05)$, although the amount of variation explained by the regression was still low $(r^2 = 0.105)$. A graph of the residuals is more evenly distributed, but still shows some similarity between successive values. A correlogram of the residuals now shows no autocorrelation at lag 1, but significant autocorrelation at lag 2 (Figure 21.11). The analysis was rerun, with the inclusion of a second term to include the annual rainfall from two years before:

$$Y_t = a + B_1 Y_{t-1} + B_2 Y_{t-2} \qquad (21.15)$$

This too was significant $(Y_t = 334.9 + 0.456Y_{t-1} - 0.412Y_{t-2}: F_{2,43} = 7.43, \ P < 0.01)$ and explains more than a quarter of the variation $(r^2 = 0.257)$. Note that for this example the constant for the first lag is positive, but the one for the second lag is negative, showing a positive relationship with the previous year's rainfall, but a negative one for the year before that.

A graph of the residuals (Figure 21.12(a)) has a more even spread of points and the correlogram of these residuals (Figure 21.12(b)) shows no low lag autocorrelation (but two significant values at higher lags). Finally, although the rainfall predicted from Equation (21.15) is not identical to actual rainfall, it is a surprisingly good predictor of years in the category "rainfall exceeds 350 mm" and is therefore useful in planning extractive operations at Neostrata 4 during the coming year (Figure 21.12(c)). This is another example of how a complex sequence can be modeled by adding additional terms to a regression equation.

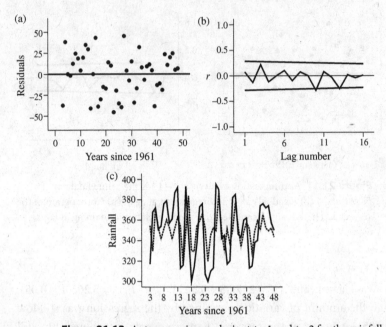

Figure 21.12 Autoregression analysis at $t-1$ and $t-2$ for the rainfall data at Neostrata 4. (a) Residuals from an autoregression at $t-1$ and $t-2$. (b) Correlogram of the residuals. The autoregression model has accounted for the autocorrelation at low lags. (c) Actual rainfall (solid line), and predicted rainfall (dashed line) from Equation (21.15).

21.10   More complex series with a cyclic component

Even a relatively complex time series can often be modeled, detrended and interpreted using autocorrelation and regression. Figure 21.13 shows sea level, measured in mm relative to a set reference point, at Port Magmago in the Western Pacific every December from 1960 to 2009. This relationship is more complex than the previous examples. The sequence does not appear to be stationary because sea level is generally lower during the middle, and higher at the beginning and end of the sequence. There is also a marked component of peaks and troughs.

A linear regression is clearly inappropriate for this non-stationary series, so quadratic and cubic polynomials were fitted in an attempt to model the general trend. Both were significant (Table 21.2) and explained about half the variation despite the superimposed cyclic component.

Table 21.2 Regression statistics for the linear, quadratic and cubic relationship fitted to the data for sea level shown in Figure 21.13. The value of r^2 is extremely low for the linear relationship and the slope of the line is not significant. In contrast, both the quadratic and cubic relationships are significant ($P < 0.001$ in each case) and both have detected a significant long-term change over time despite the cyclic component.

Equation	r^2	F	Probability	a	b_1	b_2	b_3
Linear	0.015	$F_{1,46} = 0.682$	0.413	236.1	0.777	NA	NA
Quadratic	0.499	$F_{2,45} = 22.453$	< 0.001	382.8	−16.94	0.362	NA
Cubic	0.513	$F_{3,44} = 15.477$	< 0.001	352.2	−9.59	−0.01	0.005

NA = not applicable because this term does not occur in the particular regression equation

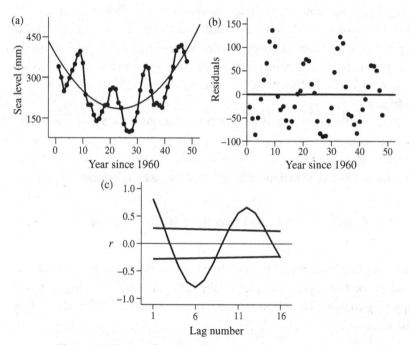

Figure 21.13 Data for sea level, in relation to a set reference point, at Port Magmago from 1960 to 2009. There appears to be a cyclic component superimposed over a trend that is not linear. (a) The quadratic line of best fit (heavy line) appears to approximate the overall trend, but not the strong cyclic component. (b) Residuals are fairly evenly distributed about the line, showing that the quadratic model has detrended the data, but there is still a cyclic pattern with time. (c) A correlogram of residuals confirms the strong cyclic pattern remaining in the detrended data.

The sums of squares for any improvement of the cubic over the quadratic, and of the quadratic over the linear model, were calculated by subtraction using the method in Table 21.1. This showed the cubic was not significantly better than the quadratic ($F_{1,44} = 1.26$, NS), which was a highly significant improvement over the linear model ($F_{1,45} = 43.59$, $P < 0.001$), so the quadratic was used.

The quadratic line of best fit in Figure 21.13(a) appears to follow the general trend. The residuals are shown in Figure 21.13(b). A regression of the residuals does not differ significantly from zero, which shows that the quadratic has detrended the data to produce a stationary plot. Nevertheless, the residuals still show a cyclic pattern and a correlogram of the residuals still shows gross apparently cyclic autocorrelation (Figure 21.13(c)).

At this stage, you might decide the analysis is sufficient and conclude that the sequence shows a significant overall trend and a fairly regular repetitive component. A more detailed regression model could include a component based on a regular pattern such as a sine wave. These models are described in more advanced texts and can be easily applied using statistical software.

Finally, although the quadratic appears to be a good fit to the overall trend in the sequence, it may not be a reliable predictor of future sea level (Figure 21.13(a)). Here too, we have deliberately used an example to emphasize the risk of extrapolating outside the measured range of a variable.

21.11   Statistical packages and time series analysis

If you are working with a relatively simple time series, the methods given here may be all you need. The availability of statistical software has made the analysis of time series easy, especially when drawing correlograms and applying complex regression models, but care is needed when interpreting the results.

21.12   Some very important limitations and cautions

A correlogram gives the correlation coefficient at numerous lags of the same sequence. Therefore, even if the data in a sequence only vary at random, for a significance level of α you would expect this proportion of values of r to fall outside the confidence interval simply by chance. For example, with a random sequence and the significance level set at 5%, you would expect 5% of values of r to be outside the 95% confidence interval and therefore be considered "significant." This is Type 1 error, as discussed in Chapter 9, and because of it **the occasional isolated significant value of r in the correlogram may mean very little.** In contrast significance at low lags, several successive significant lags, or a repeated pattern of significant positive and negative values (e.g. Figure 21.13(c)) suggests autocorrelation that is "real."

Equation (21.5) for autocorrelation given earlier in this chapter is for a relatively long sequence where the mean and the standard deviation can be treated as parameters of the population. For shorter sequences this equation needs to be modified because it may be more appropriate to treat the sequence, or the overlapping sections, as samples (Davis, 2002). For long series, the slightly different equations give very similar results.

It is extremely important to remember that the conclusions drawn are only as reliable as the data. A very short part of a sequence (e.g. years 1963–8 of Figure 21.13(a)) may show an extremely significant increase or decrease over time but may not be representative of the entire sequence. Therefore, as a general rule, the reliability of the conclusions increases with sequence length.

When assessing autocorrelation, provided **the sequence is long** (e.g. at least 40–50 observations) **and the number of calculated lags are few** (e.g. **up to no more than one quarter of the sequence length**), the correlogram should give a realistic estimate. For shorter sequences, or very high lag numbers relative to the length of the sequence, the estimate may be unrealistic. This is why most statistical programs for calculating autocorrelation stop when the lag interval exceeds a certain proportion of the sequence length.

Most software packages now include methods for time series analysis including seasonal or repeated effects and autoregressive functions. These methods need to be used and interpreted with care and you should seek expert advice if a sequence is complex.

21.13 Sequences of nominal scale data

Often a sequence consists of data measured on a nominal scale (Chapter 3), where the variable consists of numbers in discrete and mutually exclusive categories (such as different rock types). For example, you might have data for a core in which the type of rock has been recorded at regular intervals, or whenever it changed. For both sets of data, you are likely to be interested in whether each type of rock occurs at random, or if there is significant dependence, such as type A being more likely to occur after type B. A sequence can be classified as: (a) random, where no state shows any dependence on the one occurring before it, (b) partially dependent where a particular state is more likely to occur after another but will not always do so and (c) fully deterministic where a particular state always occurs after another. Partially dependent sequences are examples of **Markov chains,** where the occurrence of successive states is neither entirely random, nor absolutely deterministic.

21.13.1 Sequences that have been sampled at regular intervals

Nominal scale data that have been recorded at regular intervals can be analyzed for association using the chi-square test described in Chapter 18. For example, petroleum geologists log well cores of shallow water stratigraphic sequences in terms of whether they represent sedimentary deposits from constant (C), rising (R) or falling (F) sea level, based on sedimentation rates and whether the sediments coarsen or fine upwards and downwards. In the following example, sequence stratigraphy is given at 124 successive depth increments of one meter each.

Top of sequence: C, R, R, C, F, C, R, R, R, R, F, F, C, F, R, C, R, R, C, C, C, C, R, F, R, R, C, R, F, C, R, R, F, R, C, F, F, F, R, R, C, C, R, R, C, R, R, R, R, C, F, R, F, R, R, C, R, F, C, R, R, F, R, C, F, F, F, R, R, C, C, R, R, C, R, R, R, R, R, R, F, F, C, F, R, C, R, R, C, C, C, C, R, F, R, R, C, R, F, C, R, R, F, R, C, F, F, F, R, R, C, C, C, R, R, C, R, R, R, R, C, F, R, F, R: Bottom of sequence.

First, to establish the relative proportions of C, R and F, the number of occurrences of each are divided by the grand total. For the 124 sampling units above, there are 35 cases (0.282) of C, 61 (0.492) of R and 28 (0.226) of F. Therefore, if you were to take one sampling unit at random from within the sequence, the values in brackets are the probabilities of it being each tract type.

Although it is useful to know the probability any sampling unit chosen at random will contain a particular tract, questions such as "If I drill down into this deposit and find tract type C, what are the probabilities that the tract in the next increment down will be C or R or F?" are likely to be of more interest.

To find the conditional probabilities that each of type C, R or F is followed by C, R or F, you need to work downwards through the sequence and, for each increment, count the number of times C is followed by each of C, R and

Table 21.3 Data for the number of transitions from tract types C, R and F to types C, R and F. The probability that each tract type will be succeeded in the next increment by type C, R or F is obtained by dividing the number of transitions in each category by the row total, and is shown in brackets.

| From tract type | To tract type | | | |
	C	R	F	Row total
C	9 (0.257)	18 (0.514)	8 (0.229)	35
R	19 (0.317)	29 (0.483)	12 (0.200)	60
F	6 (0.214)	14 (0.500)	8 (0.286)	28
Column total	34	61	28	123

F; when R is followed by each of C, R and F, and when F is followed by each of C, R and F. These data are often called **transitions**, even when several successive sampling units contain the same tract type so there is no change from increment to increment. The total number of transitions will usually be one less than the number of sampling units in a sequence, because there may not be anything recorded above the top one. For the 124 sampling units in the sequence above, there are 123 transitions.

The numbers of cases where tracts C, R and F transition to C, R and F are in Table 21.3. Each **row** gives the data where a particular tract type is followed by another (C, R or F), and each **column** gives the data for cases where each tract type is preceded by another. Therefore, for the 35 cases of "from C", the next increment within the sequence is C in 9/35 cases, R in 18/35 cases and F in 8/35 cases. Dividing the numbers in each of these three categories by their row total gives the probabilities that tract type C will be followed in the next increment by type C, type R or type F. The same procedure applies to tract R and tract F, and has been done in Table 21.3.

These conditional probabilities can be used to test whether the likelihoods of finding tract types C, R and F differ from their actual proportions within the sequence. This comparison is shown in Table 21.4. The expected numbers of each transition, if the three tract types occur at random, are the number of cases of the "from" tract type multiplied by the proportion of tract type C or R or F within the entire sequence. For example, there are 35 cases of transitions "from" C, so the expected number of these followed by R is: $35 \times$ (the proportion of R in the entire sequence) = $(35 \times 0.492) = 17.2$. The expected values are in Table 21.4.

Table 21.4 Data for the number of cases where tract types C, R and F transition to types C, R and F, together with the expected numbers of each transition calculated using the proportions of each tract type in the sequence.

	To tract type			
From tract type	C	R	F	Row total
C	obs 9	obs 18	obs 8	35
	exp 9.9	exp 17.2	exp 7.9	
R	obs 19	obs 29	obs 12	60
	exp 16.9	exp 29.5	exp 13.5	
F	obs 6	obs 14	obs 8	28
	exp 7.9	exp 13.8	exp 6.3	

The data are in the form of a 3×3 contingency table (Chapter 18), from which expected values can be calculated and compared to those observed with a chi-square test, provided no more than 20% of expected frequencies are less than five (Section 18.3.1):

$$\chi^2 = \sum_{i=1}^{n} \frac{(o_i - e_i)^2}{e_i} \qquad (21.16 \text{ copied from } 18.1)$$

with the number of degrees of freedom being the product of (*from states* −1) ×(*to states* −1) which in this case is 4.

If this assumption cannot be met, a randomization test can be used (Section 18.3.3).

By inspection there is not a great deal of difference between any observed and expected value in Table 21.4 so you are unlikely to be surprised that the value of chi-square is not significant ($\chi^2_4 = 1.47$). The probability of striking a particular tract type one meter lower in the sequence is not conditional upon the type present in the preceding sample. The sequence shows only randomness, and therefore does not have Markovian properties.

21.13.2 Sequences for which only true transitions have been recorded

For sequence data where only transitions that are a **true change of state** (i.e. from one rock type to another) have been recorded, a more complex

Table 21.5 The number of true transitions from state C to R or F, state R to C or F, and state F to C or R. The matrix has vacant cells along the diagonal because a state cannot transition to itself.

From state	To state			
	C	R	F	Row total
C		18	8	26
R	19		12	31
F	6	14		20
Column total	25	32	20	77

analysis is needed. Table 21.5 gives data for the true transitions in the sequence used above. For this type of data the cells along the diagonal of the matrix are always vacant: a state cannot change to itself. These are called **embedded sequences**. Importantly, and unlike the previous example, it is **not appropriate** to treat the matrix (Table 21.5) as a contingency table (where perhaps the three vacant cells are each incorrectly given the value of zero) and there has been a great deal of debate about how to analyze such data. The recommended solution is to create a set of expected values using matrix algebra, and Davis (2002) gives a fully worked example.

An embedded sequence can also be classified as (a) random, (b) partially dependent where a particular state is more likely to occur after another but will not always do so and (c) fully deterministic where a particular state always occurs after another. Embedded sequences with significant partial dependence have Markovian properties.

21.14 Records of the repeated occurrence of an event

Another type of sequence often analyzed by earth scientists is **the repeated occurrence of the same type of event over time or distance**. For example, you might have historical data for the dates of all floods that exceeded a stream rise of 2.0 meters recorded from 1850–2009 for the Santo Mindelo River (Table 21.6) and need to know if the frequency of occurrence of flood events of this severity is changing over time.

Data for the frequency of an event can be extracted from temporal records by subdividing the list of dates of occurrence (here years) into several time intervals of equal duration and then counting the numbers of

Table 21.6 The dates of all floods > 2.0 meters, recorded from 1850–2009 for the Santo Mindelo River. Note that there was more than one flood in some years, and none in the final year 2009.

1850	1866	1887	1908	1934	1959	1977	1990
1851	1866	1890	1908	1934	1960	1978	1991
1852	1867	1892	1911	1938	1961	1978	1991
1852	1870	1894	1911	1939	1962	1979	1992
1853	1873	1895	1912	1942	1962	1980	1993
1854	1874	1895	1914	1943	1963	1983	1994
1854	1874	1896	1921	1944	1966	1983	1995
1854	1874	1900	1923	1946	1969	1984	1997
1856	1875	1903	1924	1950	1972	1984	2000
1857	1880	1904	1924	1952	1972	1986	2000
1859	1881	1904	1924	1953	1973	1987	2001
1860	1882	1905	1931	1955	1974	1987	2007
1861	1886	1907	1933	1956	1974	1988	2008

Table 21.7 The numbers of all floods > 2.0 meters within each 10-year period from 1850–2009 for the Santo Mindelo River.

Interval	Number of floods	Interval	Number of floods
1850–59	11	1930–39	6
1860–69	5	1940–49	4
1870–79	6	1950–59	6
1880–89	5	1960–69	7
1890–99	6	1970–79	9
1900–09	8	1980–89	9
1910–19	4	1990–99	8
1920–29	5	2000–09	5

the event within each. This is the same as the procedure used to construct a histogram (Section 3.3.2) and has been done for an interval length of 10 years in Table 21.7. If there is no overall change in frequency over the sequence the numbers within each interval will be similar, and a regression line of the number of events against time will be expected to have a slope of zero. The same method can be used where the data are records of distance.

Importantly, the quality of the data extracted in this way from temporal or spatial records will depend on the length of the chosen interval, just as the interval width of a histogram will affect a graphical summary of data (Section 3.3.2). Here, however, interval width is particularly important. If it is too short there may be many cases where the extracted data are zero, which will bias the regression analysis, but as it increases there will be fewer and fewer intervals, which will reduce the number of data (and degrees of freedom). You need to compromise and choose a width that preferably results in at least 80% of intervals containing one or more cases, and, if possible, have at least six intervals.

A linear regression analysis showed the sequence in Table 21.7 did not differ in slope from a line of zero ($Y = 1.071 + 0.002X : F_{1,14} = 0.044$, NS), so there is no evidence of a change in the frequency of > 2.0 meter floods of the Santo Mindelo River from 1850 to 2009 (Figure 21.14(a)). Depending on the pattern shown by the data you might also want to consider fitting a more complex regression (e.g. a quadratic). For the relationship shown in Figure 21.14(a) this does not appear necessary.

Raw data for the repeated occurrence of an event within a sequence (e.g. Table 21.6) can also be used to test whether the event is occurring at random, whether cases are clustered together, or are more evenly spaced than expected by chance. Data for the **length of the interval elapsing between successive events** can be extracted by working through a sequence such as Table 21.6. For example, the second flood in the sequence (which was recorded in 1851) occurred one year after the first record in 1850, giving an interval of one year. The third flood (1852) also occurred one year after the previous one (1851), but the fourth (also in 1852) occurred zero years after the previous. The full set of extracted data is given in Table 21.8. The shortest interval between floods is zero years, and the longest is seven, with the most frequent between-flood interval being one year (41.1% of cases).

Table 21.8 also includes the percentage of cases remaining (or **surviving**) with intervals longer than the one specified. For example, 76.7% of intervals between floods are longer than zero years, but only 8.9% are longer than three years.

These summary data can be used to produce a **survival graph** (often called an **empirical survival graph** because it has been derived from a set of empirical or "real" rather than "theoretical" data), where the X axis gives interval length from zero up to the longest (here seven years), and the Y axis gives the percentage of cases with intervals of longer duration than the one

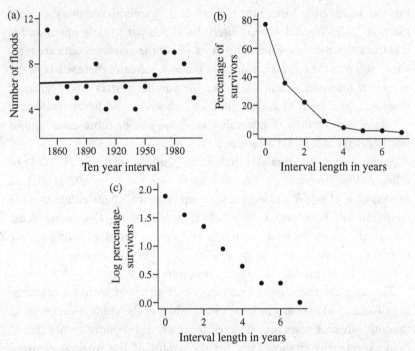

Figure 21.14 The number of floods > 2.0 meters occurring within every 10 year interval from 1850–2009 for the Santo Mindelo River. (a) The heavy line shows the line of best fit for linear regression. (b) The empirical survival function for the intervals between floods of > 2.0 meters from 1850–2009 showing the percentage of intervals longer than the duration shown on the X axis. (c) Logarithm of the percentage survivors. A line drawn through the points in (c) would be relatively straight, so the event of a flood > 2 meters appears to occur at random in time.

specified by the value of X (Figure 21.14(b)). The graph will always have a negative slope, and if the event occurs at random it should be a declining exponential. This is difficult to assess, so the logarithm of the percentage survival against interval length is usually plotted instead (Figure 21.14(c)) and should be a straight line of negative slope. For events that show pronounced clustering the empirical survival graph will initially drop steeply and then become flatter, while a more regular than expected series will give an empirical survival graph with a gentle negative slope followed by a steeper decrease at longer intervals (Figure 21.15). Incidentally, survival graphs are used in many fields, including studies of human longevity, where the use of "survivors" is particularly appropriate.

Table 21.8 The number of years elapsing between successive floods of > 2.0 meters from 1850–2009 for the Santo Mindelo River. Column 1 gives the interval elapsing between successive floods, column 2 the number of cases for each, column 3 the percentage of total cases and column 4 the percentage of cases remaining (the "survivors") having longer intervals.

Number of years elapsing between successive floods (years)	Number of cases	Percentage of total cases	Percentage of cases with longer intervals
0	21	23.3	76.7
1	37	41.1	35.6
2	12	13.3	22.2
3	12	13.3	8.9
4	4	4.4	4.4
5	2	2.2	2.2
6	0	0.0	2.2
7	2	2.2	0.0

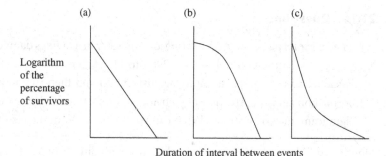

Duration of interval between events

Figure 21.15 The pattern expected for the logarithmic empirical survival function for a sequence when an event within a sequence occurs (a) at random, (b) relatively regularly and (c) in clusters.

21.15 Conclusion

The methods given here are a broad, conceptual introduction to the essentials of sequence analysis. For ratio, interval or ordinal scale sequences regression analysis is becoming increasingly popular, especially because statistical software that can run extremely complex models is now readily available. Autocorrelation is a useful method for determining the characteristics of a sequence and for subsequently testing whether a regression model is a good fit. The examples in this chapter illustrate how sequences can be modeled by iteratively refining and testing increasingly complex regression equations. We have given examples, but it is difficult to give more prescriptive methods because the regression analysis chosen will be determined by the characteristics of the sequence being investigated.

Sequences of nominal scale data can often be analyzed by extensions of the non-parametric tests described in Chapter 18, although care needs to be taken when analyzing data for true transitions. Records of repeated occurrence can be used to extract summary data from which analyses such as regression can be used to test for changes in the frequency of occurrence, and whether the event is occurring at random.

Considering the numerous pitfalls and the complexity of some methods, it is important to seek expert advice when analyzing and interpreting complex sequences. Finally, it is most important to realize that extrapolation beyond the range of a sequence may be unreliable, however good the fit of any model to the data within it.

21.16 Questions

(1) The table of data below gives the turbidity of water in a tailings dam for 48 consecutive months from January 2005 to December 2008. Use a statistical package to graph turbidity against month and then run an autocorrelation analysis. Are there any obvious seasonal trends? Are there any significant autocorrelations? What can you conclude from the results?

Month	Turbidity	Month	Turbidity	Month	Turbidity	Month	Turbidity
1	303	13	392	25	396	37	364
2	326	14	337	26	313	38	362
3	317	15	316	27	399	39	388
4	370	16	366	28	390	40	315
5	372	17	345	29	321	41	327
6	325	18	300	30	345	42	340
7	354	19	319	31	393	43	374
8	392	20	337	32	320	44	343
9	379	21	338	33	300	45	372
10	346	22	376	34	385	46	315
11	361	23	307	35	371	47	300
12	382	24	300	36	343	48	328

(2) Explain why a statistician said "Autocorrelation is very useful for relatively long sequences, but for a sequence of 20 the significance of any autocorrelation means little, especially at high lags."

(3) The data in the table below are for summers when heatwaves occurred (with a heatwave defined as a continuous sequence of five or more days on which the maximum temperature exceeded 40 °C), at Port Nundy in southern Australia from 1850 to 2009. You have been asked to test the hypothesis "The frequency of heatwaves has not changed from 1850–2009 at Port Nundy." (a) What analysis would you use? (b) What can you conclude from this analysis? (c) Do heatwaves appear to occur at random during this temporal sequence?

1850	1881	1912	1948	1974
1851	1884	1914	1950	1978
1852	1885	1919	1951	1980
1853	1890	1921	1953	1981
1854	1892	1924	1957	1982
1856	1893	1925	1958	1984
1859	1895	1926	1960	1986
1864	1899	1934	1961	1988
1865	1900	1935	1962	1990
1870	1903	1939	1964	1992
1877	1904	1944	1968	1995
1878	1907	1945	1971	2004
1880	1910	1947	1973	2009

22 | Introductory concepts of spatial analysis

22.1 Introduction

Earth scientists often rely on different types of **maps** where the **spatial location of each sampling unit** is one of the variables of interest. For example, a geoscientist might have data for the presence or absence of copper-bearing ore at 54 test cores drilled within the **sampling space** of a 10 000 square mile mining lease. The effectiveness of any further prospecting would be improved if you knew whether the spatial distribution of cores showing copper-bearing ores occurred at random or in some sort of pattern within the sampling space. The methods for summarizing and analyzing such data are called **spatial analyses.**

Even though the Earth is three-dimensional, most summary spatial information is presented as two-dimensional maps representing the Earth's surface, often with an overlay to indicate other variables. For example, maps showing landforms are printed on two-dimensional sheets, with contour lines and numbers to show the third dimension of elevation. Similarly, a map of the location and flow per minute of test wells for oil might indicate flow with numbers (e.g. the average number of barrels per day from each well) or display this as the proportional height of a single bar at each well.

The location of any point on a two-dimensional surface can be accurately and precisely defined by its X and Y coordinates, which are the distances in two directions at 90° to each other from a set reference point, in just the same way as a graph is used to display a two-dimensional scatter plot of bivariate data. Interpretation of a flat, two-dimensional display is so easy that even though the Earth is approximately spherical and any location on the surface of this sphere can be defined by the degrees, minutes and seconds of its latitude and longitude, cartographers have developed several transformations for projecting this sphere on a two-dimensional surface for use as a map.

This chapter is an introduction to some of the essential concepts of two-dimensional spatial analysis.

22.2   Testing whether a spatial distribution occurs at random

The distribution of points within a two- (or higher) dimensional sampling space can occur (a) at random, (b) more regularly than expected by chance and (c) less regularly than expected by chance. Figure 22.1 gives three examples.

One procedure used to test for randomness within two dimensions is to subdivide a sampling space into several smaller replicate blocks of equal size that are often called **quadrats** (Figure 22.2). The number of points within each quadrat is counted and used to produce a summary table of frequencies

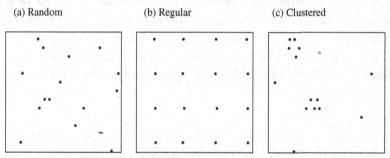

Figure 22.1 Examples of the distribution of points in two-dimensional space. (a) A random pattern. (b) A regular pattern is more uniform. (c) A clustered pattern is less uniform than random.

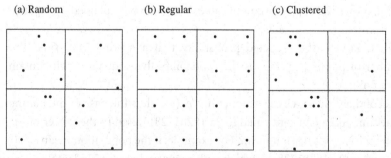

Figure 22.2 When a sampling space is subdivided into quadrats of equal size, the variation among quadrats will be (a) intermediate for a random distribution, (b) smaller for a regular one and (c) larger for a clustered one.

(Chapter 3) showing the number of quadrats containing zero, 1, 2 ,3 .... etc. points that can be used to assess the spatial distribution of the variable.

If the distribution is random, then the frequency of cases per quadrat would be expected to have a distribution and variance that would be characteristic of a random variable.

If the points are more evenly distributed than expected by chance, the variation among quadrats will be relatively small (because each quadrat will contain very similar numbers).

If the points are less evenly distributed than expected by chance (and therefore **clustered**), the variation among quadrats will be relatively large (because some will contain no or very few points, but others will contain a lot).

You can assess whether any departure from randomness is statistically significant by comparing the observed frequencies to those expected for a random variable. One of the most straightforward ways of doing this is to calculate the frequencies expected for a Poisson distribution (Chapter 7).

The mean number of points per quadrat is the total number of points (m) within the sampling space, divided by the number of quadrats (n):

$$\overline{X} = \frac{m}{n} \tag{22.1}$$

This sample mean is used as an estimate of the population mean μ in the following formula, which gives the expected proportion of quadrats containing each number of points expected under the Poisson distribution:

$$P(X) = \frac{e^{-\mu}\mu^X}{X!} \tag{22.2}$$

Equation (22.2) appears complicated, but it can be explained by separating it into its components.

First, $P(X)$ is the expected probability that a quadrat will contain X number of points (e.g. $P(3)$ is the probability that a quadrat will contain three points).

Second, $e^{-\mu}$ specifies the exponential, e, (which is the base of the natural logarithm and a constant equal to 2.718281728) raised to the power of $-\mu$, where μ is the population mean. For example, if the population mean is 2.0, then $e^{-\mu} = (2.718281728)^{-2}$, which is the square root of 2.718281728.

Third, μ^X is the population mean raised to the power of X, where X is the expected number of points per quadrat. For example, for a population mean of 2.0 and the expected category of three points per quadrat, $\mu^X = 2^3$, which is 8.

Finally, $X!$ is the symbol for the **factorial** of X and specifies "All of the positive integers up to and including X multiplied together and then multiplied by 1.0." For example:

$$3! = (1 \times 2 \times 3) \times 1 = 6$$
$$5! = (1 \times 2 \times 3 \times 4 \times 5) \times 1 = 120$$

Importantly:

$$1! = (1) \times 1 = 1$$

and

$$0! = (\text{no numerical value}) \times 1 \text{ which also equals } 1.$$

Note here that "no numerical value" does not mean zero. It means that no number is present (because there are no positive integers less than zero).

Using Equation (22.2) for a population mean of 3.0, the expected Poisson probability of a quadrat containing five points is:

$$P(5) = \frac{(2.71828)^{-3} \times (3.0)^5}{5!}$$

which is 0.01. So if the distribution of points is random, 1% of the total number of quadrats would be expected to contain five points. The same formula can be used to calculate the expected proportions of quadrats containing any number of points.

Equation (22.2) is tedious to calculate, but Box 22.1 gives all the functions you need to write a Microsoft Excel® spreadsheet that will automatically give the expected Poisson probabilities for any number per quadrat, provided the mean is known.

One analysis for non-randomness uses the chi-square test (Chapter 18) to compare the observed and expected frequencies. There is a worked example in Section 22.2.1.

22.2.1 A worked example

If you ever want to find a meteorite, go to Antarctica! The combination of the lack of ground cover (making meteorites easy to spot) and flow of ice

Box 22.1 A spreadsheet Poisson calculator

The following table can be written into a Microsoft Excel® spreadsheet to give a calculator for the expected Poisson probability for quadrats containing zero, 1, 2 3 ... etc. points, provided the mean is known. It uses the formula:

$$P(X) = \frac{e^{-\mu}\mu^X}{X!} \qquad \text{(22.3 copied from 22.2)}$$

Open a new blank Microsoft Excel® spreadsheet. Start at cell A1 (the top left hand cell in the spreadsheet) and write in the following descriptions and formulae *exactly* as they are given below. You must start at cell A1, in which you should write the word 'Mean'. The calculator must only occupy the block of cells from A1 to B6. Make sure you include the minus sign in the worksheet function given in cell B3, which must be written as: =EXP(-B1).

	A	B
1	Mean	
2	Expected number per quadrat	
3		=EXP(-B1)
4		=POWER(B1,B2)
5		=FACT(B2)
6	Probability	=(B3*B4)/B5

Once you have entered all the functions given above, type the numerical value of the mean into cell B1 and the number per quadrat (here you might start with zero) in cell B2 immediately below this. Do not type numbers into any other cells. The spreadsheet will automatically give you the probability for each number per quadrat if the Poisson distribution applies. This probability is displayed to the right of "Probability," in cell B6 where you had specified the worksheet function =(B3*B4)/B5.

As a quick check, if you put in a mean of 3.805 in cell B1 and an expected number per quadrat of 4.0 in cell B2, the probability should be 0.19440924. Your spreadsheet should look the same, or similar to the one here (depending on the number of decimal places you specify for the cells):

	A	B
1	Mean	3.805
2	Expected number per quadrat	4
3		0.0222592
4		209.613208
5		24
6	Probability	0.19440924

Table 22.1 The frequencies of quadrats containing zero, 1, 2, etc. meteorites, and the expected frequencies if the spatial distribution of meteorites is random.

Number of meteorites per quadrat	Observed frequency of cases	Expected proportion if random	Expected frequency if random
0	10	0.022592	4.4
1	14	0.084696	16.9
2	9	0.161135	32.2
3	23	0.204372	40.9
4	65	0.194409	38.9
5	74	0.147945	29.6
6	5	0.093822	18.8
7	0	0.050999	10.2
8 and more	0	0.040621	8.1
Total	$n = 200$	1.000000	
Mean	3.805		

against mountains (which brings old ice to the surface, where it is deflated by winds that expose meteorites) makes Antarctica a premier hunting ground for meteorite recovery. Every year since 1976, at least one team of scientists has spent several weeks during the southern hemisphere summer searching for meteorites from stranding surfaces along the Transantarctic Mountains of Antarctica. Their job would be made easier if the distribution of meteorites could be better understood.

Table 22.1 gives summary data for the number of quadrats containing zero, 1, 2, etc. meteorites within 200 km^2 that has been subdivided into 200 quadrats of equal area. Note that the expected frequencies have been combined for more than 7 meteorites per quadrat. In total, there were 761 meteorites distributed within the 200 quadrats.

The expected proportions of quadrats containing each number of meteorites were calculated using Equation (22.2) and multiplied by 200 to give the expected frequencies. Finally the observed and expected frequencies were used in Equation (22.4) to obtain the chi-square statistic for the nine (0 – 8+ meteorites per quadrat) categories:

$$\chi^2 = \sum_{i=1}^{n} \frac{(o_i - e_i)^2}{e_i} \qquad \text{(22.4 copied from 18.1)}$$

Box 22.2 The number of degrees of freedom for a chi-square comparison when the formula for the expected frequencies includes a sample statistic

The value of chi-square for the comparison of k observed and expected frequencies derived from the Poisson distribution has $k - 2$ degrees of freedom. In Chapter 18 and elsewhere, however, it was explained that the number of degrees of freedom for a goodness-of-fit test is $k - 1$. For a fixed total of observations, the numbers within $k - 1$ categories are free to vary, but those in the "final" category to be filled can only be one number, which is therefore a fixed quantity.

When expected frequencies are derived externally (e.g. an expectation of 3 : 1 on the basis of some hypothesized property of the system, or 1 : 1 : 1 on the basis of an expectation of equal frequencies, as discussed in Chapter 18), the degrees of freedom are one less than the number of categories (Section 18.2).

In example 22.2.1, the expected proportions per category have been calculated using the Poisson formula (22.2), which **used the sample mean as the best estimate of the population mean**. The use of a statistic taken from the sample being tested will result in the loss of one more degree of freedom, because for a set sample mean, all but one of the values of the sampling units contributing to that mean are free to vary. Therefore, the degrees of freedom for the chi-square test in this example must be $k - 2$.

If instead you have an independent estimate of the population mean (perhaps from a more extensive study), the number of degrees of freedom for the chi-square test is $k - 1$.

---

which is: $\chi^2_7 = 144.82$, $P < 0.001$. The distribution of meteorites is significantly non-random.

If a chi-square analysis shows the distribution is non-random, you are very likely to want to know whether it is more even, or more clustered. For a Poisson distribution, the variance and the mean are numerically the same. For a regular distribution, the variance will be smaller than the mean and for a clustered distribution it will be larger.

The mean number per quadrat is the total number of points (here the 761 meteorites) divided by the number of quadrats (here 200), giving a mean of 3.805:

$$\overline{X} = \frac{m}{n} \qquad\qquad \text{(22.5 copied from 22.1)}$$

The variance is calculated using the formula for the sample variance:

$$s^2 = \frac{\sum\limits_{i=1}^{n} (X_i - \overline{X})^2}{n - 1} \qquad\qquad \text{(22.6 copied from 7.6)}$$

where n is the number of quadrats and X_i is the number of meteorites in the first, through to the final quadrat (here the 200th).

The **variance to mean ratio** is the variance divided by the mean:

$$\frac{s^2}{\overline{X}} \qquad\qquad \text{(22.7)}$$

and should be about 1.0 for a random distribution.

The value of $s^2/\overline{X}$ can be compared to an expected value of 1.0 by using a single-sample t test, although a modification has to be made to the formula for that test given in Chapter 8. The variance to mean ratio is derived from the sample variance, but the formula for the t test includes the standard error of the mean, which is **also** calculated from the sample variance. To use both in the same formula will result in bias. Instead, an independent (and conservatively large) **estimate** of the standard error of the mean is made only from n, the number of quadrats:

$$\text{SEM}(est) = \sqrt{\frac{2}{n - 1}} \qquad\qquad \text{(22.8)}$$

Therefore the equation for the single-sample t test is:

$$t_{n-1} = \frac{\frac{s^2}{\overline{X}} - 1.00}{\text{SEM}(est)} \qquad\qquad \text{(22.9)}$$

where n is the number of quadrats. If $s^2/\overline{X}$ is 1.0, the numerator will be zero and so will the value of t. A significant value means that the distribution is either even or clustered: when the variance divided by the mean is smaller than 1.0, the distribution is more even and when it is greater than 1.0, the distribution is more clustered. Sometimes this statistic is given as the mean divided by the variance, so you need to make sure which version has been used when interpreting results in reports and publications.

For the example in Table 22.1, the sample variance is 2.17 and the mean is 3.805. Therefore the variance to mean ratio is 0.57. The independent estimate of the SEM is:

$$\sqrt{\frac{2}{n-1}} = \sqrt{\frac{2}{199}} = 0.100251.$$

When these values are put into Equation (22.9):

$$t_{199} = \frac{0.57 - 1.00}{0.100251} = -4.29$$

which is highly significant (Appendix A). The distribution of meteorites is relatively even among the 200 quadrats. The significant result is consistent with the procedure for comparing the distribution to the Poisson described earlier in this section, but because the variance to mean test is relatively conservative you may find in some cases that the "Poisson" method is just significant at $P < 0.05$ and the less powerful variance to mean ratio test is not. In these cases the variance to mean ratio still indicates the direction (e.g. even or clustered) of the departure from a random distribution.

22.2.2 The problem of scale

The procedure for assessing randomness described above is sensitive to quadrat size. For example, a distribution may show extreme clustering on a relatively small scale that may not be detected on a larger scale (Figure 22.3). This is another example of the usefulness of making a preliminary inspection of a set of data, which could be done by plotting the distribution of points on a two-dimensional map and examining it for any obvious pattern (Figure 22.3).

22.2.3 Nearest neighbor analysis

Another test for a pattern in spatial data that is not sensitive to the problem of scale described above, is **nearest neighbor analysis**. This is used in many areas of science, including studies of the spatial distributions of plants and animals. As the number of points within a sampling space

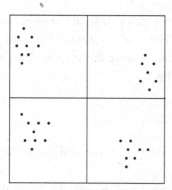

Figure 22.3 A relatively large quadrat size may not detect clustering on a smaller scale. The number of points within each of the four quadrats is very similar (thus suggesting an even distribution) despite extreme clustering.

Box 22.3 A Monte Carlo approach to assessing randomness: the spatial distribution of impact craters on Venus

Another way of assessing a spatial pattern for randomness is to use the **Monte Carlo method** (Chapter 18) to take a large number of simulated random samples and create a distribution that would apply under the null hypothesis.

Photographs from the Magellan mission show that the planet Venus has relatively few impact craters and these appear to be randomly distributed. The low number of craters suggests the surface is relatively young, because craters accumulate over time: for example the Moon has older terrain that is heavily cratered (like the light-colored highlands) and younger terrain with significantly fewer craters (like the basaltic maria). The random distribution of craters suggests there is little or no recent tectonic activity, because this will continually create patches of new crust where there will initially be no craters, thereby giving a spatial distribution of craters that is clustered rather than random. The hypothesis that the surface of Venus is relatively young with little or no recent tectonic activity was highly controversial. Strom *et al.* (1994) compared the observed spatial distribution of impact craters on Venus with that expected for a random distribution, using a Monte Carlo simulation, and concluded that the distribution of craters was random. The implied lack of recent tectonic activity on Venus is now widely accepted.

increases, the average distance between each point and the one closest to it (its nearest neighbor) will decrease. Here too, a formula derived from the Poisson is used to predict the **expected average distance between nearest neighbors** when the pattern is random:

$$\bar{\delta} = \frac{1}{2}\sqrt{\frac{A}{n}} \qquad (22.10)$$

where A is the area of the sampling space and n is the total number of points within it.

This expected value can be compared to the observed mean distance between nearest neighbors, $\bar{d}$, from the sample. The standard error of the mean (SEM) of the nearest neighbor distances is:

$$\text{SEM} = \sqrt{\frac{(4-\pi)A}{4\pi n^2}} \qquad (22.11)$$

using the constant π (3.147) and A and n as in the previous equation.

The Z test is used to compare $\bar{d}$ and $\bar{\delta}$:

$$Z = \frac{\bar{d} - \bar{\delta}}{\text{SEM}} \qquad (22.12)$$

Once Z is outside the range of ± 1.96, the mean distance between nearest neighbors is significantly different than that expected if the points are distributed at random.

Although nearest neighbor analysis is not affected by scale, it is sensitive to **edge effects** that occur because points close to the edge of the sampling space can only have neighbors within it. If the space were larger, the points might have closer nearest neighbors (Figure 22.4).

One solution to this bias is to designate a strip around the outer edge of the sampling space as a **guard region** and only measure the nearest neighbor distances **from** each of the points within the remaining (reduced) area in the middle (Figure 22.5). Some of the nearest neighbors of these points may be in the guard region, and the distances measured out to them will be realistic if the guard region is wide enough. Unfortunately, however, using part of the sample space as a guard region reduces the sample size.

Another method is to create an **artificial guard region** by making eight copies of the sampling space and placing these around the original, as shown in Figure 22.6. Here too, the nearest neighbor distances are only measured

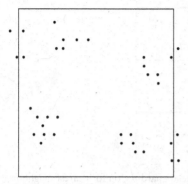

Figure 22.4 Nearest neighbor analysis can be subject to edge effects. Some of the points within the sampling space have "true" nearest neighbors that are outside the boundary, but their nearest neighbors within the sampling space are further away. Restricting the measurements to nearest neighbors within the sampling space will overestimate the mean nearest neighbor distance.

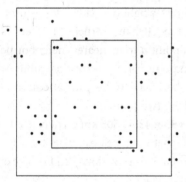

Figure 22.5 Use of a guard region to reduce edge effects. Nearest neighbor distances are only measured from points within the central (and therefore reduced) sampling space. Some nearest neighbors to these points may be within the guard region, and the distances to them will be representative provided the guard region is wide enough.

from points within the original sampling space. This method does not reduce the sample size.

22.2.4   A worked example

Table 22.2 gives the nearest neighbor distances for the 17 points within the 35×43 km central region shown in Figure 22.5. Points have been numbered in order, running across and down from the top left-hand corner. It is

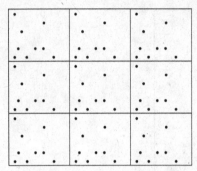

Figure 22.6 Use of an artificial guard region for nearest neighbor analysis. The original sample space (center square) has been copied eight times, and the copies placed in the same orientation around it. Nearest neighbor distances are only measured from points within the central sampling space.

important to note that the nearest neighbor distance is measured from **every point** within the sampling space. Therefore, for two points designated as "*a*" and "*b*" that are each other's nearest neighbors within the central region (e.g. points 1 and 2 just inside the top left-hand corner of the central region of Figure 22.7), the distance from point *a* to its nearest neighbor point *b* is 2.7 km, but so is the distance from point *b* to its nearest neighbor, which in this case is point *a*. For *n* points there have to be *n* distances, so the same distance is recorded for each of these two points.

The area of the central region is $35 \times 43 = 1505\,\text{km}^2$. There are 17 points, so the **expected** average distance between nearest neighbors (Equation (22.10)) is 4.705 km, with an SEM of 0.596 (Equation (22.11)). The observed mean distance, $\bar{d}$, from Table 22.2 is 4.662.

The Z score (Equation (22.12)) is: $\frac{4.662-4.705}{0.596} = -0.072$. This is within the range of ± 1.96, so the mean nearest neighbor distance is no different to that expected if the distribution is random.

22.3   Data for the direction of objects

An object with a recognizable direction that has a "head and a tail" (i.e. it represents an arrow rather than a line) within a two-dimensional (or higher) sampling space will face in a particular **direction**. For two dimensions, this can be quantified by a single number in relation to a reference (e.g. true North, the true vertical or some defined direction) in degrees of a circle from 1° to 360°. Summary statistics for directional data have many

Table 22.2 The distances, in kilometers, to their nearest neighbor, for the 17 numbered points within the central area of Figure 22.7.

Point number	Distance	Point number	Distance
1	1.65	10	2.70
2	1.65	11	2.70
3	3.40	12	3.95
4	3.40	13	4.75
5	5.90	14	4.25
6	8.30	15	3.90
7	8.30	16	3.90
8	8.30	17	3.90
9	8.30	**Average:**	4.662

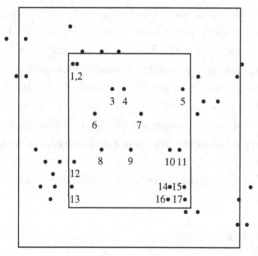

Figure 22.7 For the 17 points (numbered in order running across and down from the top left-hand corner within the central region of the sampling space), the nearest neighbor distances for points 1 and 2 are the same, because each is the other's nearest neighbor. The same applies to points 3 and 4; 10 and 11; 15 and 17. The nearest neighbors of points 5, 12 and 13 are in the guard region. The central area is 35 × 43 km.

applications. For example, the directions of streams in a floodplain can be used to model the effects of rainfall upon a catchment and historical data for the directions moved by cyclones and hurricanes can be used to predict their paths.

22.3.1 Summarizing and displaying directional data

If you have data for the wind or water flow directions at several different locations, they can be summarized and displayed by an extension of the method used to generate a histogram (Chapter 3 and Section 22.2). For each location, the direction is recorded in degrees, from one to 360°. Next the circle is subdivided into several equal arcs (e.g. 1–90°, 91–180°, 181–270° and 271–360°), and the number of cases within each is counted to give a table of frequencies. These divisions are often called **bins**. The summary data could be displayed as a conventional histogram (Chapter 3), but to show the actual or relative directions of the objects they are usually plotted in a **circular histogram**, which is a circle divided into several equal arcs, equivalent to the bars of a conventional histogram, with the radius of the filled area indicating the frequency within each bin. This is called a **rose diagram** because a circle subdivided into filled arcs of different radii somewhat resembles a flower. Two examples are shown in Figure 22.8. Here, just as for a conventional histogram, the number of bins and their 'width' in degrees must be chosen to give a meaningful display (Chapter 3).

Rose diagrams can be visually misleading, because the width and area of an arc increase with distance from the origin. Therefore, if only the **petal length** is **proportional to the frequency of cases within each arc**, the perceived importance of relatively low counts will be reduced and that of relatively high ones increased (Figure 22.8(a)). For this reason, rose diagrams are usually plotted with **the area of each petal being proportional to the frequency of cases** (Figure 22.8(b)).

22.3.2 Drawing a rose diagram

The following method gives a rose diagram where the largest frequency always extends to the maximum radius of the rose. For example, for a rose diagram of radius 24 mm, the angular division containing the largest count will have this petal length.

To draw a rose diagram where the length of each petal is proportional to its frequency, the number of cases within each bin is counted from the raw data. Petal lengths are calculated using the formula:

$$\text{petal length} = \frac{r_{max} \times \text{Freq}_{pet}}{\text{Freq}_{max}} \tag{22.13}$$

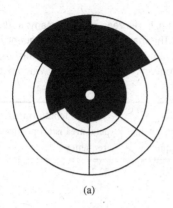

(a)

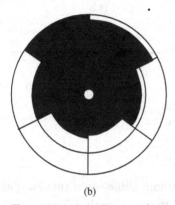

(b)

Figure 22.8 Rose diagrams from the data in Table 22.3. (a) Petal length is directly proportional to the frequency within each angular division. (b) Petal length when petal area is directly proportional to the frequency within each division. Note that (a) gives the visual impression of a far less symmetrical distribution.

where r_{max} is the maximum radius of the rose diagram, Freq_{pet} is the frequency for a particular petal and Freq_{max} is the highest frequency within the set of petals. Therefore, for the highest frequency, Freq_{pet} will be equal to Freq_{max} so its petal length will be r_{max}. There is worked example in Section 22.3.3.

To draw a rose diagram where the area of each petal is proportional to its frequency, the procedure described above is followed, except that petal

Table 22.3 Summary data for the direction, in degrees, of 250 streams in the Channel country of western Queensland. The maximum observed frequency is 73 and the maximum radius of the rose diagram has been chosen as 24 mm. (a) Petal lengths that are directly proportional to frequency. (b) Petal lengths, when petal area is directly proportional to frequency.

Direction	Observed frequency	(a) Petal length (mm) when proportional to frequency	(b) Petal length (mm) when petal area is proportional to frequency
1–60	63	20.7	22.3
61–120	31	10.2	15.6
121–180	20	6.6	12.6
181–240	22	7.2	13.2
241–300	41	13.5	18.0
301–360	73	24.0	24.0
Total	250		

length is calculated using the formula:

$$\text{petal length} = \sqrt{\frac{r^2_{max} \times \text{Freq}_{pet}}{\text{Freq}_{max}}} \tag{22.14}$$

where r^2_{max} is the square of the maximum radius of the rose diagram. Here too, when Freq_{pet} is equal to Freq_{max}, the petal length will be r_{max}. There is a worked example below.

22.3.3 Worked examples of rose diagrams

Data for the directions of 250 streams in the Channel country of western Queensland, as summarized for six equal angular divisions of 60°, are given in Table 22.3. These have been used to calculate petal length when it is directly proportional to frequency (Equation (22.13)), and petal length when the petal area is directly proportional to frequency (Equation (22.14)). The two rose diagrams are shown in Figure 22.8. By inspection, the diagram where petal length is proportional to frequency gives the misleading perception that a greater proportion of the objects face towards the upper part of the rose than does the one where petal area is proportional to frequency. When interpreting rose diagrams, it is important to know which scale has been used! It is not unknown for mining companies to produce prospectuses containing misleading rose diagrams. The use of the "area proportional" type is recommended.

22.3.4 Testing whether directional data show a pattern

There are numerous tests for whether the directions faced by a sample of objects shows a relatively even angular distribution or some pattern. If all directions are equally likely, then the expected number of observations within each equal subdivision should, on average, be the same, and the rose diagram will be symmetrical. This hypothesis can be tested by a chi-square goodness of fit between the observed and expected frequencies:

$$\chi^2 = \sum_{i=1}^{n} \frac{(o_i - e_i)^2}{e_i} \qquad (22.15 \text{ copied from } 18.1)$$

The number of degrees of freedom is one less than the number of bins (see Box 22.2).

22.3.5 A worked example

Table 22.4 gives summary data for the direction, in degrees, for the sample of 250 streams in Table 22.3, together with the expected frequency when there are equal numbers in every bin. The chi-square goodness of fit test is significant: $\chi^2_5 = 57.77$, so $P < 0.001$, indicating that the orientation of the valleys is not equally likely in all directions. By inspection, it appears that most are facing from the north-west through to the north-east ($301°$ through $0°$ to $60°$).

22.3.6 Data for the orientation of objects

Objects without a definable "head or tail" (e.g. where the object being mapped is a line or a plane) cannot be assigned a specific direction in a two-dimensional space. They face in **two directions**, which for a straight object (such as the strike of contacts between overlying formations) will be $180°$ apart (e.g. $24°$ and $204°$). Instead, data for the **orientations** of these objects are often given as two directions and both are recorded in the table of frequencies used to draw a rose diagram (Section 22.3.2). For straight objects, the rose diagram showing orientation will always be perfectly symmetrical, with petals

Table 22.4 Summary data, for the direction, in degrees, of the 250 streams in Table 22.3 with the expected number within each bin if the objects sampled are equally likely to face in any direction.

Direction	Observed frequencies	Expected
1–60	63	41.66
61–120	31	41.66
121–180	20	41.66
181–240	22	41.66
241–300	41	41.66
301–360	73	41.66

of identical length occurring opposite to each other. **Caution is needed** when using these data to test hypotheses about whether objects are orientated at random because the table of frequencies will contain $2n$ data from n objects and therefore inflate the value of a statistic such as chi-square. One solution for straight objects is simply to double all the angles (Krumbein, 1939). For a circle subdivided into 360°, this always gives the same angle for both directions (e.g. an object orientated at 30° and 210° will have doubled angles of 60° and 420°, the latter of which is 360° + 60° and thus equal to 60°). So you need to record the value of the doubled angle only **once** in a table of frequencies used for a chi-square test for randomness (Section 22.3.5).

22.4 Prediction and interpolation in two dimensions

Often data are obtained for a ratio or interval scale variable that is measured at several locations, but there is a need to predict its value elsewhere within the same sampling space. For example, you might have data for the percentage of nickel at ten locations within a mining lease and want to predict the areas where additional exploration is likely to discover high yielding deposits within that lease.

One very important property of spatial data can be used to make such predictions. Often a variable such as the depth of a water table or the thickness of a coal deposit shows **regional dependence**, which means that its value at two or more locations close to each other is **relatively similar**. For example, the percentage yield of a deposit of gold-bearing rock might be 1.1 g/ton and 1.2 g/ton at two sites only 100 meters apart, but 1.1 and 5.6 g/ton at sites 1200 meters apart. This is just the same as positive

Figure 22.9 Illustration of regional dependence. The depth of shading indicates the similarity between the value of a variable at the central point (X_1) decreasing with distance away from it (X_2), until there is no dependence (X_3).

autocorrelation at low lags (Chapter 21), where values for points close to each other within a sequence are more similar than those further apart. In two dimensions, the space within which regional dependence occurs can be visualized as a circle, with the amount of dependence decreasing with distance outwards from the central point (Figure 22.9). This can be extremely helpful in estimating the value of a variable at sites relatively close to those where it is known.

22.4.1   The semivariance and semivariogram

A statistic that quantifies the amount of regional dependence between two points is the **semivariance**:

$$\gamma = \frac{(X_i - X_j)^2}{2} \tag{22.16}$$

where X is the value of the variable at points X_i and X_j. For two identical values the semivariance will be zero. As the difference between them increases, so will the semivariance, which can only ever be zero or greater.

The semivariance is the same as the **variance** for a sample containing only two points. For any sample from a population the variance is:

$$s^2 = \frac{\sum_{i=1}^{n}(X_i - \overline{X})^2}{n - 1} \tag{22.17 copied from 7.6}$$

For a sample of only two (X_1 and X_2), the denominator ($n - 1$) is always 1.0, so Equation (22.17) for the variance becomes:

$$(X_1 - \overline{X})^2 + (X_2 - \overline{X})^2 \tag{22.18}$$

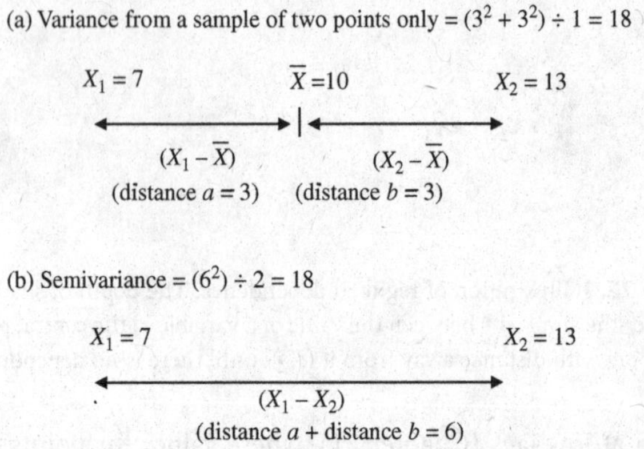

(a) Variance from a sample of two points only = $(3^2 + 3^2) \div 1 = 18$

$X_1 = 7$ $\qquad\qquad$ $\overline{X} = 10$ $\qquad\qquad$ $X_2 = 13$

$\longleftrightarrow$ | $\longleftrightarrow$

$\qquad$ $(X_1 - \overline{X})$ $\qquad\qquad$ $(X_2 - \overline{X})$
$\qquad$ (distance $a = 3$) $\qquad$ (distance $b = 3$)

(b) Semivariance = $(6^2) \div 2 = 18$

$X_1 = 7$ $\qquad\qquad\qquad\qquad\qquad\qquad$ $X_2 = 13$

$\longleftrightarrow$

$\qquad\qquad\qquad$ $(X_1 - X_2)$
$\qquad\qquad$ (distance a + distance $b = 6$)

Figure 22.10 Graphical explanation of why the variance for a sample of two points is the same as the semivariance. (a) Each value is equidistant from the mean and the variance is therefore (distance $a)^2$ + (distance $b)^2$. (b) This is the same as the semivariance (distance a + distance $b)^2 \div 2$ because mathematically, if $a = b$, $a^2 + b^2 = \frac{(a+b)^2}{2}$.

where $\overline{X}$ is the mean of the two values. This is the same as the semi-variance (22.16), because the sum of the two squared differences $(X_1 - \overline{X})^2 + (X_2 - \overline{X})^2$ is mathematically equal to the **difference** between the points squared and divided by two: $(X_1 - X_2)^2/2$. A graphical explanation of this equation is shown in Figure 22.10. If you were to take only two points at random from a population, Equations (22.16) and (22.17) will each estimate the population variance (but neither is likely to give an accurate estimate because the sample size is only two).

The importance of the semivariance is its use **as an accurate and precise statistic** to quantify the **dissimilarity** of a variable between a specifically chosen central point (X_1) and each of several other points $(X_2, X_3,$ etc.) increasingly distant from it. For each pair of points $((X_1X_2), (X_1X_3),$ etc.), the value of the semivariance is plotted on the Y axis against the distance between them on the X axis, to give a scatter plot called the **experimental** (or sometimes the **empirical**) **semivariogram**. The relationship between the semivariance and distance from the central point will depend on the amount of regional dependence.

If there is **no regional dependence**, then the value of the variable at the central point will be unrelated to its value elsewhere. So the scatter plot of the semivariance will simply display a range of values, each of which is an estimate of the population variance from a sample where $n = 2$ (Figure 22.11(a)).

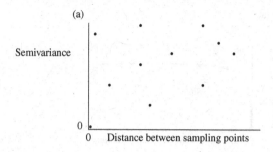

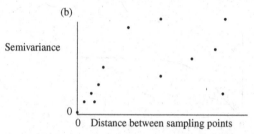

Figure 22.11 The experimental semivariogram is a scatter plot of the semivariance against distance between sampling points, where $X = 0$ is the central location. (a) No regional dependence. (b) Strong regional dependence, shown by the small semivariances between the central location and those nearby.

If there **is some regional dependence**, then the values of the variable at the central point and those nearby will be similar, thereby giving relatively small semivariances. As the distance from the central point increases, the amount of dependence reduces, so the semivariances will tend to increase but also become more scattered (Figure 22.11(b)). At this distance (and beyond), the two points are equivalent to having been chosen at random from the population, so each of the widely scattered semivariances will estimate the population variance for samples of $n = 2$.

Once the experimental semivariogram has been plotted, a smoothed line of best fit called the **theoretical semivariogram** is fitted through the points with the restriction that it must start from a relatively low value at the central point, subsequently increase, but eventually plateau out. Theoretical semivariograms have been fitted to the two scatter plots in Figure 22.12.

When there is no regional dependence, the theoretical semivariogram will rise extremely rapidly to a plateau that is equal to the population

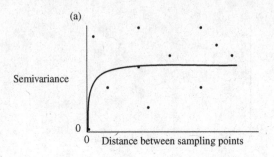

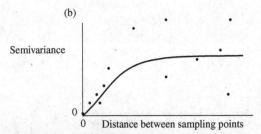

Figure 22.12 The theoretical semivariogram is a smoothed line fitted to the scatter plot of the semivariances against distance between sampling points, where $X = 0$ is the central location. (a) No regional dependence. (b) Strong regional dependence.

variance (Figure 22.12(a)). It will be a relatively good estimate of this parameter because it is the **average** of several semivariances scattered around it.

When there is regional dependence, the theoretical semivariogram will initially have a low value near the central point, subsequently increase, but eventually plateau out when the central point and those more distant from it are no longer related. Here too, the averaged value at the plateau gives a relatively good estimate of the population variance.

It may seem logical that the semivariance for two replicates taken at the central location (and therefore "no distance" apart) should be zero (e.g. Figure 22.12). However, this does not necessarily occur because there may be **within-site variation** (which is the same as the "error" discussed in Chapter 10) on an extremely small spatial scale that will give a relatively small minimum semivariance (Figure 22.13).

The features of the theoretical semivariogram are shown in Figure 22.14. The semivariance at $X = 0$ is called the **nugget** or **nugget effect**. When the

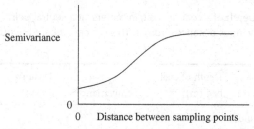

Figure 22.13 Variation among replicates taken at the same location will give a semivariance of more than zero at the central point.

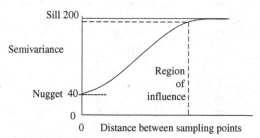

Figure 22.14 Features of the theoretical semiovariogram. When $X = 0$ the value of the semivariance is called the nugget. The maximum value (at the plateau) is the sill. The region of influence is the value of X (estimated graphically by reverse prediction) for which the theoretical semivariance is 95% of the distance between the sill and the nugget.

semivariance reaches its maximum height at the plateau, its value is called the **sill**. The outer limit of the **region of influence** surrounding the central point is defined as the value of X when the semivariance has reached 95% of the **difference between the sill and the nugget**. For example, the distance between a nugget of 40 and a sill of 200 is 160, so the value of the semivariance at the outer limit of the region of influence is $40 + (0.95 \times 160) = 192$. The region of influence can be estimated graphically by reverse prediction as shown in Figure 22.14 and represents the distance outwards from the central location within which the variable shows some regional dependence.

The theoretical semivariogram used to be fitted by eye, but there are several equations available to estimate it. One of the most commonly used is the **exponential**:

$$Y = c + (S - c) \times (1 - e^{-\frac{3h}{a}}) \tag{22.19}$$

Table 22.5 Depth of a coal deposit, in meters, at a central point and 15 other pits at various distances from it. This is a case where the nugget is zero.

Pit number	Depth of coal bed (m)	Semivariance	Distance from Pit 1 (km)
1 (central point)	11	0.0	0
2	12	0.5	1
3	11	0.0	3
4	14	4.5	4
5	34	264.5	14
6	4	24.5	20
7	42	480.0	28
8	32	220.5	32
9	17	18.0	40
10	28	144.5	54
11	34	265.5	67
12	25	98.0	70
13	37	338.0	76
14	41	450.0	79
15	7	8.0	100
16	29	162.0	125

where Y is the semivariance, c is the nugget, S is the sill, e is the natural logarithm, h is the distance from the central point and a is the range of influence. This will always give a relationship that increases but eventually plateaus out.

22.4.2 A worked example

Table 22.5 gives data for the depth, in meters, of a West Virginia coal deposit at a central location and for 15 pits increasingly distant from it.

The semivariances were calculated using Equation (22.16). The experimental semivariogram is shown in Figure 22.15 and suggests some regional dependence because the semivariances between Pit 1 and each of Pits 2, 3 and 4 are relatively small. Simply by inspection, the theoretical semivariogram has a nugget of zero, a sill of about 220 km, and a region of influence of about 25 km, with the last obtained by graphical reverse prediction from a semivariance of $0 + 0.95 \times (220 - 0)$ (see above).

(a)

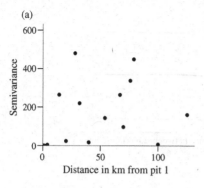

Distance in km from pit 1

(b)

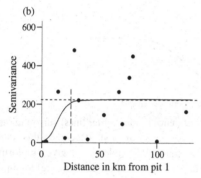

Distance in km from pit 1

Figure 22.15 (a) Experimental semivariogram showing the semivariance plotted against the distance in kilometers from the central point of Pit number 1. (b) Theoretical semivariogram for an exponential function fitted to the same data. The horizontal dashed line shows the sill. The vertical dashed line shows the region of influence of about 25 km, estimated graphically from a semivariance of 209 that was calculated as 95% of the difference between the sill and the nugget.

22.4.3 Application of the theoretical semivariogram

One important application of the theoretical semivariogram is to predict the value of a variable at sites where it has not been measured. The width of the 95% confidence interval around the line of the theoretical semivariogram will depend on the amount of regional dependence. When there is **no regional dependence**, the line will rise rapidly and the 95% confidence interval around it will be relatively wide, because it is the smoothed average of many estimates made when $n = 2$. When there **is regional dependence**,

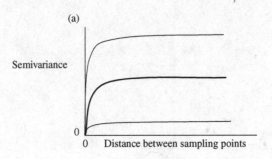

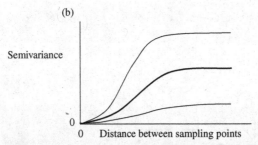

Figure 22.16 The 95% confidence limits (lighter lines) surrounding the line
for the theoretical semivariance (heavy line) will depend on the extent of
regional dependence. (a) No regional dependence will give a uniform
confidence interval (and thus a very imprecise semivariance) for any point
outside the central one. (b) When there is regional dependence, the confidence
interval will initially be narrow but will increase (thus initially showing high
precision, which will decrease with distance from the central point).

the line will rise more slowly. Its 95% confidence interval will initially be
very narrow because the regional dependence surrounding the central point
will constrain the estimates of the semivariance to within a relatively small
range (Figure 22.16).

If a variable shows regional dependence and the point(s) at which you
want to predict it lie within the regions of influence of known locations, it is
possible to make quite precise estimates of its value. This is the basis of the
method of interpolation called **Kriging** (named after D. G. Krige who
developed the technique).

This is an extreme simplification, but essentially Kriging gives the solution to
a set of simultaneous equations so that the value at an unknown site is the best

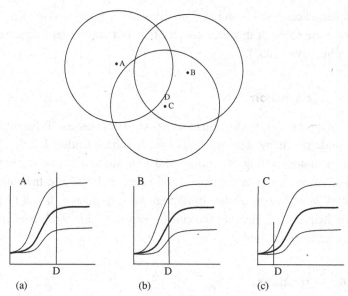

Figure 22.17 The value of a variable is known for sites A, B and C, and needs to be estimated for site D. The region of influence is shown as a circle around each known site. Note that point D is near the edge of the region of influence for A, closer to the central point of B, and very close to the central point of C. The value of the variable estimated for D must be one that gives the best possible fit to its position on the three theoretical semivariograms (darker lines), taking into account the 95% confidence intervals (lighter lines), for A, B and C.

possible overall fit to the appropriate points on the theoretical semivariograms that overlap it. A graphical example to illustrate the concept is given in Figure 22.17. The value of the variable is not known for point D, but D lies within the regions of influence of points A, B and C where the variable is known. Note that D is closest to the central region of C and therefore positioned on the left-hand side of the theoretical semivariogram for C, where the semi-variance is relatively low (Figure 22.17(c)). Consequently, C has the greatest influence upon the estimate of the value of D, which must be close to the actual value of C at the central point in order to fit within the 95% confidence interval of the theoretical semivariogram. In contrast, B has less influence (Figure 22.17 (b)) and A has the least influence on D (Figure 22.17(a)).

An enormous amount has been written (including many advanced text-books) on Kriging, together with computer programs for carrying out the complex calculations required. The detailed methods used for Kriging are beyond the scope of this introductory text, but an excellent explanation is given by Davis (2002).

22.5   Conclusion

This chapter is an introduction to some essential concepts of spatial analysis that underpin many of the methods used by earth scientists. It is designed to give an understanding of random and non-random spatial distributions, and an introduction to the analysis of directional data and the concept of estimation and interpolation using regional dependence. It will be particularly helpful in evaluating the conclusions in reports and from research that uses these methods.

22.6   Questions

(1) The table below gives the observed number of silver mines in 200 quadrat samples, each $1 \, \text{km}^2$ in area, in northern Idaho. The total number of mines was 276 and the mean number per quadrat is 1.38.

Number of mines per quadrat	Observed frequency of cases
0	44
1	87
2	31
3	27
4	9
5	2
6	0
7	0
8 and more	0
Total	$n = 200$
Mean	1.38

(a) Calculate the expected proportion, and the expected frequency, of quadrats containing zero, 1, 2, etc. mines if the distribution is random and test the hypothesis that the observed distribution is random using the methods in Section 22.2.1. Is the value of chi-square significant? (b) What is the variance and the variance to mean ratio of the number of mines per quadrat? (c) Is this ratio significantly different from 1.0? (d) Is the distribution of silver mines per quadrat more or less uniform than random?

(2) Behavior patterns of dinosaurs may be recorded in the geologic record in serendipitous circumstances, such as the many sets of tracks exposed in what is now the Connecticut River Valley in the eastern US. A paleontologist wanted to test the idea that an exposure of 32 different dinosaur footprints in a large slab of sandstone show the movements of a herd, pack, or flock of dinosaurs all moving in the same direction. The following data are for the direction, in degrees, of 32 fossil footprints: 101°, 134°, 205°, 95°, 164°, 199°, 223°, 144°, 173°, 243°, 105°, 164°, 266°, 119°, 146°, 169°, 94°, 155°, 250°, 144°, 170°, 201°, 215°, 99°, 134°, 219°, 232°, 140°, 178°, 232°, 107° and 161°. (a) Use these data to plot a rose diagram, where petal area is proportional to frequency, showing the direction of the footprints. (b) By inspection of this diagram, do the directions of the footprints appear to be evenly distributed? (c) Test this hypothesis.

(3) A homeowner decided to plant a line of trees along the edge of her lot to give some shelter from winds and storms. She needed to know the most frequent direction (if any) of the wind in order to maximize the protection provided by the wind break. The following data are for the direction, in degrees, of weekly average wind directions recorded over the course of a year at a nearby airport: 2°, 164°, 272°, 82°, 330°, 200°, 30°, 164°, 123°, 180°, 224°, 302°, 44°, 12°, 294°, 316°, 205°, 164°, 150°, 242°, 107°, 64°, 33°, 348°, 12°, 124°, 242°, 182°, 350°, 241°, 50°, 134°, 163°, 70°, 204°, 332°, 144°, 112°, 94°, 16°, 305°, 264°, 50°, 142°, 127°, 164°, 233°, 48°, 154°, 293°, 210° and 48°. (a) Use these data to plot a rose diagram, where petal area is proportional to frequency, showing the direction of the winds. (b) By inspection of this diagram do the wind directions appear to be evenly distributed? (c) Test this hypothesis.

(4) What is the difference between the theoretical and the empirical semivariogram?

(5) Comment on the statement "If there is no regional dependence the semivariogram has little or no use in interpolation such as Kriging."

23 | Choosing a test

23.1   Introduction

Statisticians and earth scientists who teach statistics are often visited in their offices by a researcher or student they may never even have met before, who is clutching a dauntingly thick pile of paper and perhaps a couple of flash drives or CDs with labels like "Experiment 1" or "Trial 2". The visitor sits down, drops everything heavily on the desk and says, "Here are my results. What stats do I need?"

This is not a good thing to do. First, the person whose advice you are asking may not have the time to work out exactly what you have done, so they may give you bad advice. Second, the answer can be a very nasty surprise like "There are problems with your experimental design."

The decision about the appropriate statistical analysis needs to be made by considering the hypothesis being tested, the experimental design and the type of data. It can save a lot of time, trouble and disappointment if you think about possible ways of analyzing the data **before the sampling is done or the experiment designed**, rather than only after the data have been collected.

Tables 23.1–23.12 are a guide to choosing an appropriate analysis from the ones discussed in this book. You need to start at Table 23.1, which initially gives five columns that are mutually exclusive alternatives. Once you have decided among these, work downwards within the column you have chosen. There may be more choices and again you need to select the appropriate column and continue downwards. Eventually you will be referred to another table with more choices that lead to suggested analyses.

Table 23.1 Are you: (a) testing the hypothesis that one or more **samples of univariate, ratio, interval ordinal or nominal scale data are from the same population**, (b) testing whether **two variables are related**, (c) examining whether there is **similarity or dissimilarity among samples of multivariate data**, (d) analyzing and/or modeling a **temporal or spatial sequence**, or (e) analyzing the spatial properties of a **distribution of points in a two-dimensional sampling space**?

(a) Testing the hypothesis that one or more univariate samples of ratio, interval, ordinal or nominal scale data are from the same population		(b) Testing whether two variables are related	(c) Examining whether there is similarity or dissimilarity among samples of multivariate data	(d) Analyzing and/or modeling a temporal or spatial sequence	(e) Analyzing the spatial properties of a distribution of points in a two-dimensional sampling space
For ratio, interval or ordinal scale data	**For nominal scale data**	Data must be bivariate: measured on a ratio, interval or ordinal scale and can be displayed as a two-dimensional scatter plot.	**Go to Table 23.10**	**Go to Table 23.11**	**Go to Table 23.12**

For ratio, interval or ordinal scale data

For example:

Comparing the mean of a sample with an expected value of 0.025 wt%:

wt% gold
0.021
0.014
0.016
0.007
0.016
0.010

or comparing the means of **two or more samples** (e.g. length of clam fossils at three different outcrops):

Site J	Site K	Site L
6	13	28
3	17	26
1	20	23
4	16	26

Go to Table 23.2

For nominal scale data

Nominal scale data are numbers in two or more mutually exclusive categories such as the number of accumulating vs. ablating glaciers (two categories) for a sample of 200:

Accumulation	Ablation
107	93

or the numbers in two or more mutually exclusive categories (e.g. accumulation vs. ablation at several different places in the Alps):

	Site 1	Site 2	Site 3
Accumulation	32	21	36
Ablation	14	23	17

Go to Table 23.4

(b) Testing whether two variables are related

Case	X	Y
1	2	7
2	4	11
3	6	13
4	8	15
5	10	18
6	12	19
7	14	22
8	16	23

Go to Table 23.9

Table 23.2 Is the hypothesis about: (a) whether one sample is from a population with a known distribution and thus known population statistics, (b) whether two or more samples are from the same population?

One sample of ratio, interval or ordinal scale data is from a population with known (or expected) statistics	Two or more samples of ratio, interval or ordinal scale data are from the same population	
	Are the samples independent?	Are the samples related?

One sample of ratio, interval or ordinal scale data is from a population with known (or expected) statistics

Case	Data
1	165
2	150
3	159
4	145
5	170

Go to Table 23.3

Are the samples independent?

Case	Sample 1	Sample 2	Sample 3
1	X		
2	X		
3		X	
4		X	
5			X
6			X

Can you assume the data are normally distributed?

Go to Table 23.5

Are the data grossly non-normal?

Go to Table 23.6

Are the samples related?

Case	Sample 1	Sample 2	Sample 3
1	X	X	X
2	X	X	X
3	X	X	X
4	X	X	X
5	X	X	X
6	X	X	X

Can you assume the data are normally distributed?

Got to Table 23.7

Are the data grossly non-normal?

Go to Table 23.8

Table 23.3 Tests for one sample of ratio, interval or ordinal scale data.

Is the sample from a population that appears normal, or not grossly non-normal?	Is the sample from a population that is grossly non-normal (e.g. bimodal?)
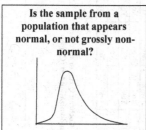	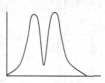
Z test if the population mean and variance are known (Chapter 8) but **single-sample *t* test** if there is only an expected value for the population mean (Chapter 8)	**Kolmogorov–Smirnov one-sample test** (Chapter 19)

Table 23.4 Tests for nominal scale data.

To test whether one sample of nominal scale data is from a population with known statistics	To test whether two or more samples of nominal scale data are from the same population			
	Are the samples independent?		**Are the samples related?**	

To test whether one sample of nominal scale data is from a population with known statistics

Are the frequencies in two or more mutually exclusive categories and you want to compare them to an expected ratio? (For example, comparison of the sex of 110 *Pryctodon* fossils to an expected ratio of 1:1).

Male	54
Female	56

Non-parametric goodness-of-fit test (Chapter 18)

Are the samples independent?

Can the data be cast as a 2×2 table?

	Sample 1	Sample 2
Male	12	33
Female	42	27

A Fisher Exact Test or a chi-square test for a 2×2 contingency table (Chapter 18)

Caution: For small sample sizes a Fisher Exact Test is recommended

---

Can the data be cast as a table with two or more columns and rows? (For example, the numbers of meteorites in the three mutually exclusive categories of stone, iron and stony iron, in three different samples).

	Sample 1	Sample 2	Sample 3
Stone	23	14	24
Iron	12	19	14
Stony iron	15	12	13

A chi-square test, exact test or randomization test for a contingency table (Chapter 18)

Caution: Exact tests are recommended when samples sizes are small

---

Are the samples related?

Are there two samples?

Case	Before treatment	After treatment
1	Yes	No
2	Yes	No
3	No	No
4	Yes	Yes
5	No	No
6	Yes	No

A McNemar test for the significance of changes or an exact or randomization test (Chapter 18)

Are there more than two samples?

Case	Condition 1	Condition 2	Condition 3
1	Yes	No	No
2	Yes	No	Yes
3	No	No	No
4	Yes	Yes	Yes
5	No	No	Yes
6	Yes	No	No

A Cochran Q test or an exact or randomization test (Chapter 18)

Table 23.5 Tests for two or more independent samples of normally distributed ratio, interval or ordinal scale data.

To test whether two or more independent samples of normally distributed ratio, interval or ordinal scale data are from the same population

---

Are the data replicates for two or more levels of one factor (e.g. Rb content of three different size fractions)?

<45 µm	45–125µm	125–250 µm
6	13	28
3	17	26
1	20	23
4	16	26

Two samples

- Sample variances similar → **Independent sample t test (Chapter 8) or single-factor ANOVA (Chapter 10)**
- Sample variances grossly dissimilar → **Independent sample t test (Chapter 8) which does not assume equal variances**

Three or more samples

- Sample variances similar → **Single-factor ANOVA (Chapter 10)**
- Sample variances grossly dissimilar → **Transform and recheck. If similar, use single-factor ANOVA (Chapter 10). If not, analyze with caution because of the increased risk of Type 1 error (Chapter 13)**

Are the data an orthogonal design for two or more factors (e.g. precipitated crystal mass in relation to temperature and cooling rate)?

	20 °C		30 °C		40 °C	
	Slow cooling	Fast cooling	Slow cooling	Fast cooling	Slow cooling	Fast cooling
6		1	13	3	28	5
3		3	17	5	26	7
1		2	20	2	23	3
4		1	16	5	26	4

Data are replicated within each combination of treatments

- Variances are similar among treatment combinations → **Two-factor ANOVA (Chapter 12)**
- Variances are grossly dissimilar among treatment combinations → **Transform and recheck variances. If similar, use two-factor ANOVA (Chapter 12). Otherwise, use ANOVA but be aware of the increased risk of Type 1 error (Chapter 13)**

Data are not replicated → **Two-factor ANOVA without replication (Chapter 14)**

Are the data an orthogonal design for two or more levels of three or more factors (e.g. stream discharge in relation to slope, velocity and vegetation)?

Three-factor ANOVA subject to the cautions and conditions for two-factor ANOVA in the columns to the left of this one

---

Are the data a nested design where one or more factors are nested within the levels of another? (For example, the effect of temperature on crystal precipitation, with two growth chambers nested within each temperature)

20°C		30°C		40°C	
Chamber A	Chamber B	Chamber C	Chamber D	Chamber E	Chamber F
6	5	13	15	28	34
3	6	17	19	26	19
2	1	20	13	23	26
1	1	16	17	26	28

Nested ANOVA (Chapter 14), subject to the cautions about inequality of variances noted in the columns for single-factor and two-factor ANOVA to the left of this one

Table 23.6 Tests for two or more independent samples of ratio, interval or ordinal scale data that are **not** normally distributed.

To test whether two or more independent samples of ratio, interval or ordinal scale data, that are grossly non-normal, are from the same population			
Are the data for two levels or samples of one factor (e.g. garnet in rocks from two different metamorphic grades, or zircon content from two outcrops)?		Are the data for three or more levels or samples of one factor (e.g. quartz content of sediments of three sizes, or trace fossils at three or more outcrops)?	
Outcrop 1	Outcrop 2	boulders cobbles pebbles	
6	13	6 13 28	
3	17	3 17 26	
1	20	1 20 23	
4	16	4 16 26	
Sample distributions similar	Sample distributions grossly different	Sample distributions similar	Sample distributions grossly different
Mann–Whitney *U* test, randomization test or exact test (Chapter 19)	**Transform to a nominal scale and analyze as categorical data** (Chapter 19)	**Kruskal–Wallis test, randomization test, or exact test** (Chapter 19)	**Transform to a nominal scale and analyze as categorical data** (Chapter 19)

Table 23.7 Tests for two or more related samples of normally distributed ratio, interval or ordinal scale data.

To test whether two or more related samples of normally distributed data are from the same population						
Two related samples			Three or more related samples			
Case	Sample 1	Sample 2	Case	Sample 1	Sample 2	Sample 3
1	12	15	1	12	15	23
2	16	12	2	16	12	18
3	21	17	3	21	17	26
4	18	10	4	18	10	21
5	19	14	5	19	14	29
6	12	18	6	12	18	24
Paired-sample *t* test (Chapter 8) or two-factor ANOVA without replication (Chapter 14)			Two-factor ANOVA without replication (Chapter 14)			

Table 23.8 Tests for two or more related samples of ratio, interval or ordinal scale data that are **not** normally distributed.

To test whether two or more related samples of data, that are grossly non-normal, are from the same population						
Two related samples			Three or more related samples			
Case	Sample 1	Sample 2	Case	Sample 1	Sample 2	Sample 3
1	12	15	1	12	15	23
2	16	12	2	16	12	18
3	21	17	3	21	17	26
4	18	10	4	18	10	21
5	19	14	5	19	14	29
6	12	18	6	12	18	24
Wilcoxon paired-sample test, randomization test or exact test (Chapter 19)			Friedman test, randomization test or exact test (Chapter 19)			

Table 23.9 Tests for whether two variables are related.

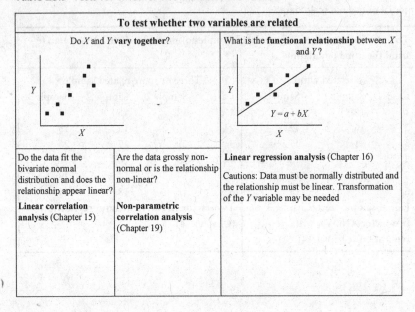

To test whether two variables are related	
Do *X* and *Y* **vary together**?	What is the **functional relationship** between *X* and *Y*?
Do the data fit the bivariate normal distribution and does the relationship appear linear? **Linear correlation analysis** (Chapter 15) · Are the data grossly non-normal or is the relationship non-linear? **Non-parametric correlation analysis** (Chapter 19)	**Linear regression analysis** (Chapter 16) Cautions: Data must be normally distributed and the relationship must be linear. Transformation of the *Y* variable may be needed

Table 23.10 Methods for comparing two or more samples for which multivariate data are available.

To isolate variables or combinations of variables that can help explain any pattern of similarity or dissimilarity among a group of multivariate samples	To display multivariate samples in a reduced two-dimensional space in order to give an easily interpreted summary of their similarities and dissimilarities	
Principal components analysis (Chapter 20)	Display as points in two-dimensional space **Multidimensional scaling** (Chapter 20)	Display as a dendogram **Cluster analysis** (Chapter 20)

Table 23.11 Methods for analyzing and modeling a temporal sequence.

Ratio, interval or ordinal scale data		Nominal scale data	
Testing whether there is repetition or relatedness among different parts of a sequence	Modeling a sequence to **explain** within sequence variability	Testing whether there is **repetition** or relatedness among different parts of a sequence	Testing whether **there is a pattern in the** repeated **occurrences** of an event
Autocorrelation (Chapter 21)	**Regression modeling of sequences** (Chapter 21)	Methods for **analyzing transitions** (Chapter 21)	**Regression analysis and survival functions** (Chapter 21)

Table 23.12 Analyzing data for a pattern related to the spatial location of sampling units occurring in a two dimensional landscape.

Testing for departures from a random pattern in two-dimensional space	Testing hypotheses about the direction or orientation of a sample of objects	Interpolation and prediction in two dimensions
Assessing whether the distribution is consistent with a random spatial pattern, using data for **counts within quadrats** (Chapter 22) Assessing whether the distribution is consistent with a random spatial pattern using **nearest neighbor analysis** (Chapter 22)	Tests for the **direction or orientation of a sample of objects** (Chapter 22)	**Calculation of the semivariance** for use in Kriging (the latter is only conceptually introduced in this text) (Chapter 22)

Appendix A: Critical values of chi-square, *t* and *F*

Table A1 Critical values of chi-square when $a = 0.05$, for 1–120 degrees of freedom. If the calculated value of chi-square is larger than the critical value for the appropriate number of degrees of freedom then the probability of the result is < 0.05 (and is therefore considered significant with an a of 0.05). For example, for three degrees of freedom the critical value is 7.815, so a chi-square larger than this indicates $P < 0.05$. Values were calculated using the method given by Zelen and Severo (1964).

Table A1

Degrees of freedom	$\alpha = 0.05$	Degrees of freedom	$\alpha = 0.05$	Degrees of freedom	$\alpha = 0.05$
1	3.841	41	56.942	81	103.010
2	5.991	42	58.124	82	104.139
3	7.815	43	59.304	83	105.267
4	9.488	44	60.481	84	106.395
5	11.070	45	61.656	85	107.522
6	12.592	46	62.830	86	108.648
7	14.067	47	64.001	87	109.773
8	15.507	48	65.171	88	110.898
9	16.919	49	66.339	89	112.022
10	18.307	50	67.505	90	113.145
11	19.675	51	68.669	91	114.268
12	21.026	52	69.832	92	115.390
13	22.362	53	70.993	93	116.511
14	23.685	54	72.153	94	117.632
15	24.996	55	73.311	95	118.752
16	26.296	56	74.468	96	119.871
17	27.587	57	75.624	97	120.990
18	28.869	58	76.778	98	122.108
19	30.114	59	77.931	99	123.225
20	31.401	60	79.082	100	124.342
21	32.671	61	80.232	101	125.458
22	33.924	62	81.381	102	126.574
23	35.172	63	82.529	103	127.689
24	36.415	64	83.675	104	128.804
25	37.652	65	84.821	105	129.918
26	38.885	66	85.965	106	131.031
27	40.113	67	87.108	107	132.144
28	41.337	68	88.250	108	133.257
29	42.557	69	89.391	109	134.369
30	43.773	70	90.531	110	135.480
31	44.985	71	91.670	111	136.591
32	46.914	72	92.808	112	137.701
33	47.400	73	93.945	113	138.811
34	48.602	74	95.081	114	139.921
35	49.802	75	96.217	115	141.030
36	50.998	76	97.351	116	142.138
37	52.192	77	98.484	117	143.246
38	53.384	78	99.617	118	144.354
39	54.572	79	100.749	119	145.461
40	55.758	80	101.879	120	146.567

Table A2 Critical two- and one-tailed values of Student's *t* statistic when α = 0.05, calculated using the method given by Zelen and Severo (1964). A *t* test is used for comparison between two samples or a sample and a population, so both **non-directional** and **directional** alternate hypotheses are possible (e.g. for the latter the alternate hypothesis might be "The mean of Sample A is expected to be greater than the population mean μ").

For non-directional and therefore two-tailed alternate hypotheses, if the calculated
   value of *t* is **outside** the range of zero ± the critical value then the probability of that
   result is < 0.05 (and therefore considered significant with an α of 0.05). For example,
   for six degrees of freedom the value of *t* must be outside the range of zero ± 2.447.

**For directional and therefore one-tailed alternate hypotheses you first need to check
   whether the difference between two means is in the direction specified by the
   alternate hypothesis** (e.g. if the hypothesis specifies mean A is greater than mean B,
   there is no point in looking up the critical value if mean B is greater than mean A,
   because the null hypothesis will stand whatever the value of *t*).

If the difference **is in the direction specified by the alternate hypothesis** then the
   absolute value of *t* (that is, the value of *t* written as a positive number irrespective of
   whether the calculated value is positive or negative) is significant for $\alpha = 0.05$ if it is
   larger than the one-tailed critical value for the appropriate number of degrees of
   freedom in Table A2. For example, for 20 degrees of freedom the absolute value of *t*
   must exceed 1.725 for $P < 0.05$.

Table A2

Degrees of freedom	$\alpha\,(2) = 0.05$	$\alpha\,(1) = 0.05$	Degrees of freedom	$\alpha\,(2) = 0.05$	$\alpha\,(1) = 0.05$
1	12.706	6.314	42	2.018	1.682
2	4.303	2.920	44	2.015	1.680
3	3.182	2.353	46	2.013	1.679
4	2.776	2.132	48	2.011	1.677
5	2.571	2.015	50	2.009	1.676
6	2.447	1.934	52	2.007	1.675
7	2.365	1.895	54	2.005	1.674
8	2.306	1.860	56	2.003	1.673
9	2.262	1.833	58	2.002	1.672
10	2.228	1.812	60	2.000	1.671
11	2.201	1.796	62	1.999	1.670
12	2.179	1.782	64	1.998	1.669
13	2.160	1.771	66	1.997	1.668
14	2.145	1.761	68	1.995	1.668
15	2.131	1.753	70	1.994	1.667
16	2.120	1.746	72	1.993	1.666
17	2.110	1.740	74	1.993	1.666
18	2.101	1.734	76	1.992	1.665
19	2.093	1.729	78	1.991	1.665
20	2.086	1.725	80	1.990	1.664
21	2.080	1.721	82	1.989	1.664
22	2.074	1.717	84	1.989	1.663
23	2.069	1.714	86	1.988	1.663
24	2.064	1.711	88	1.987	1.662
25	2.060	1.708	90	1.987	1.662
26	2.056	1.706	92	1.986	1.662
27	2.052	1.703	94	1.986	1.661
28	2.048	1.701	96	1.985	1.661
29	2.045	1.699	98	1.984	1.661
30	2.042	1.697	100	1.984	1.660
31	2.040	1.696	200	1.972	1.653
32	2.037	1.694	300	1.968	1.650
33	2.035	1.692	400	1.966	1.649
34	2.032	1.691	500	1.965	1.648
35	2.030	1.690	600	1.964	1.647
36	2.028	1.688	700	1.963	1.647
37	2.026	1.687	800	1.963	1.647
38	2.024	1.686	900	1.963	1.647
39	2.023	1.685	1000	1.962	1.646
40	2.021	1.684	∞	1.960	1.6455

Table A3 Critical values of the F distribution for ANOVA when $\alpha = 0.05$. If the calculated value of F is larger than the critical value for the appropriate degrees of freedom then it indicates a probability of <0.05 (and is therefore considered significant with an α of 0.05). The columns in Table A3 give the degrees of freedom for the numerator of the F ratio and the rows give the degrees of freedom for the denominator. For example, the critical value for $F_{3,7}$ is 4.35, so an F statistic larger than this is considered significant at $P < 0.05$. Values were calculated using the method given by Zelen and Severo (1964).

Denominator degrees of freedom	\multicolumn{18}{c}{Numerator degrees of freedom}																	
	1	2	3	4	5	6	7	8	9	10	12	15	20	30	50	100	200	254
1	161.45	199.50	215.71	224.58	230.16	233.99	236.77	238.88	240.54	241.88	243.90	245.00	248.01	250.10	252	253	254	254
2	18.51	19.00	19.16	19.25	19.30	19.33	19.35	19.37	19.38	19.40	19.41	19.43	19.45	19.46	19.47	19.49	19.49	19.49
3	10.13	9.55	9.28	9.12	9.01	8.94	8.89	8.85	8.81	8.79	8.75	8.70	8.66	8.62	8.58	8.55	8.54	8.54
4	7.71	6.94	6.59	6.39	6.26	6.16	6.09	6.04	6.00	5.96	5.91	5.86	5.80	5.75	5.70	5.66	5.65	5.65
5	6.61	5.79	5.41	5.19	5.05	4.95	4.88	4.82	4.77	4.74	4.68	4.62	4.56	4.50	4.44	4.41	4.39	4.39
6	5.99	5.14	4.76	4.53	4.39	4.28	4.21	4.15	4.10	4.06	4.00	3.94	3.87	3.81	3.75	3.71	3.69	3.69
7	5.59	4.74	4.35	4.12	3.97	3.87	3.79	3.73	3.68	3.64	3.58	3.51	3.45	3.38	3.32	3.27	3.25	3.25
8	5.32	4.46	4.07	3.84	3.69	3.58	3.50	3.44	3.39	3.35	3.28	3.22	3.15	3.08	3.02	2.97	2.95	2.95
9	5.12	4.26	3.86	3.63	3.48	3.37	3.29	3.23	3.18	3.14	3.07	3.01	2.94	2.86	2.80	2.76	2.73	2.73
10	4.96	4.10	3.71	3.48	3.33	3.22	3.14	3.07	3.02	2.98	2.91	2.85	2.77	2.70	2.64	2.59	2.56	2.56
11	4.84	3.98	3.59	3.36	3.20	3.10	3.01	2.95	2.90	2.85	2.79	2.72	2.65	2.57	2.51	2.46	2.43	2.43
12	4.75	3.89	3.49	3.26	3.11	3.00	2.91	2.85	2.80	2.75	2.69	2.62	2.54	2.47	2.40	2.35	2.32	2.32
13	4.67	3.81	3.41	3.18	3.03	2.92	2.83	2.77	2.71	2.67	2.60	2.53	2.46	2.38	2.31	2.26	2.23	2.23
14	4.60	3.74	3.34	3.11	2.96	2.85	2.76	2.70	2.65	2.60	2.53	2.46	2.39	2.31	2.24	2.19	2.16	2.16
15	4.54	3.68	3.29	3.06	2.90	2.79	2.71	2.64	2.59	2.54	2.48	2.40	2.33	2.25	2.18	2.12	2.10	2.10
16	4.49	3.63	3.24	3.01	2.85	2.74	2.66	2.59	2.54	2.49	2.43	2.35	2.28	2.19	2.12	2.07	2.04	2.04
17	4.45	3.59	3.20	2.97	2.81	2.70	2.61	2.55	2.49	2.45	2.38	2.31	2.23	2.15	2.08	2.02	1.99	1.99
18	4.41	3.56	3.16	2.93	2.77	2.66	2.58	2.51	2.46	2.41	2.34	2.27	2.19	2.11	2.04	1.98	1.95	1.95
19	4.38	3.52	3.13	2.90	2.74	2.63	2.54	2.48	2.42	2.38	2.31	2.23	2.16	2.07	2.00	1.94	1.91	1.91

20	4.35	3.49	3.10	2.87	2.71	2.60	2.51	2.45	2.39	2.35	2.28	2.20	2.12	2.04	1.97	1.91	1.88
21	4.32	3.47	3.07	2.84	2.69	2.57	2.49	2.42	2.37	2.32	2.25	2.18	2.10	2.01	1.94	1.88	1.84
22	4.30	3.44	3.05	2.82	2.66	2.55	2.46	2.40	2.34	2.30	2.23	2.15	2.07	1.98	1.91	1.85	1.82
23	4.28	3.42	3.03	2.80	2.64	2.53	2.44	2.38	2.32	2.28	2.20	2.13	2.05	1.96	1.88	1.82	1.79
24	4.26	3.40	3.01	2.78	2.62	2.51	2.42	2.36	2.30	2.26	2.18	2.11	2.03	1.94	1.84	1.80	1.77
25	4.24	3.39	2.99	2.76	2.60	2.49	2.40	2.34	2.28	2.24	2.16	2.09	2.01	1.92	1.84	1.78	1.75
26	4.23	3.37	2.98	2.74	2.59	2.47	2.39	2.32	2.27	2.22	2.15	2.07	1.99	1.90	1.82	1.76	1.73
27	4.21	3.35	2.96	2.73	2.57	2.46	2.37	2.31	2.25	2.20	2.13	2.06	1.97	1.88	1.81	1.74	1.71
28	4.20	3.34	2.95	2.71	2.56	2.45	2.36	2.29	2.24	2.19	2.12	2.04	1.96	1.87	1.79	1.73	1.69
29	4.18	3.33	2.93	2.70	2.55	2.43	2.35	2.28	2.22	2.18	2.10	2.03	1.94	1.85	1.77	1.71	1.67
30	4.17	3.32	2.92	2.69	2.53	2.42	2.33	2.27	2.21	2.17	2.09	2.01	1.93	1.84	1.76	1.70	1.66
40	4.08	3.23	2.84	2.61	2.45	2.34	2.25	2.18	2.12	2.08	2.00	1.92	1.84	1.74	1.66	1.59	1.55
50	4.03	3.18	2.79	2.56	2.40	2.29	2.20	2.13	2.07	2.03	1.95	1.87	1.78	1.69	1.60	1.52	1.48
60	4.00	3.15	2.76	2.53	2.37	2.25	2.17	2.10	2.04	1.99	1.92	1.84	1.75	1.65	1.56	1.48	1.44
70	3.98	3.13	2.74	2.50	2.35	2.23	2.14	2.07	2.02	1.97	1.89	1.81	1.72	1.62	1.53	1.45	1.40
80	3.96	3.11	2.72	2.49	2.33	2.21	2.13	2.06	2.00	1.95	1.88	1.79	1.70	1.60	1.51	1.43	1.38
90	3.95	3.10	2.71	2.47	2.32	2.20	2.11	2.04	1.99	1.94	1.86	1.78	1.69	1.59	1.49	1.41	1.36
100	3.94	3.09	2.70	2.46	2.31	2.19	2.10	2.03	1.97	1.93	1.85	1.77	1.68	1.57	1.48	1.39	1.34
200	3.89	3.04	2.65	2.42	2.26	2.14	2.06	1.98	1.93	1.88	1.80	1.72	1.62	1.52	1.41	1.32	1.26
500	3.86	3.01	2.62	2.39	2.23	2.12	2.03	1.96	1.90	1.85	1.77	1.69	1.59	1.48	1.38	1.28	1.21

Appendix B: Answers to questions

2.8 (1) The "hypothetico-deductive" model is that science is done by proposing a hypothesis, which is an idea about a phenomenon or process that may or may not be true. The hypothesis is used to generate predictions that can be tested by doing a mensurative or a manipulative experiment. If the results of the experiment are consistent with the predictions the hypothesis is retained. If they are not (for an experiment that appears to be a good test of the predictions) the hypothesis is rejected. By convention, a hypothesis is stated as two alternatives: the null hypothesis of no effect or no difference, and the alternate hypothesis which states an effect. For example, "Apatite treatment affects the amount of lead leached from soil" is an alternate hypothesis, and the null is "Apatite treatment does not affect the amount of lead leached from soil." Importantly a hypothesis can never be proven because there is always the possibility that new evidence may be found to disprove it. A "negative" outcome, where the alternate hypothesis is rejected, is still progress in our understanding of the natural world and therefore just as important as a "positive" outcome where the null hypothesis is rejected.

2.8 (2) The value recorded from only one sampling or experimental unit may not be very representative of the remainder of the population.

4.9 (1) An example of confusing a correlation with causality is when two variables are related (that is, they vary together) but neither causes the other to change. For example, as depth in the ocean increases, light intensity decreases and pressure increases, but the decrease in light intensity does not cause the increased pressure or vice versa.

4.9 (2) "Apparent replication" is when an experiment (either mensurative or manipulative) contains replicates, but the placement or collective treatment of the replicates reduces the true amount of replication. For example, if you had two different treatments replicated several times within only two furnaces set at different temperatures the level of replication is actually the furnace in each treatment (and therefore one). Another example could be two different heavy metal rehabilitation treatments applied to each of 10 plots, but all 10 plots of one treatment were clustered together in one place on a mining lease and all 10 of the other treatment were clustered in another.

5.6 (1) Copying the mark for an assignment and using it to represent an examination mark is grossly dishonest. First, the two types of assessment are different. Second, the lecturer admitted the variation between the assignment and exam mark was "give or take 15%" so the relationship between the two marks is not very precise and may severely disadvantage some students. Third, there is no reason why the relationship between the assignment and exam mark observed in past classes will necessarily apply in the future. Fourth, the students are being misled: their performance in the exam is being ignored.

5.6 (2) It is not necessarily true that a result with a small number of replicates will be the same if the number of replicates is increased, because a small number is often not representative of the population. Furthermore, to claim that a larger number was used is dishonest.

6.11 (1) Many scientists would be uneasy about a probability of 0.06 for the result of a statistical test because this non-significant outcome is very close to the generally accepted significance level of 0.05. It would be helpful to repeat the experiment.

6.11 (2) Type 1 error is the probability of rejecting the null hypothesis when it is true. Type 2 error is the probability of rejecting the alternate hypothesis when it is true.

6.11 (3) The 0.05 level is the commonly agreed upon probability used for significance testing: if the outcome of an experiment has a probability of less than 0.05 the null hypothesis is rejected. The 0.01 probability level is sometimes used when the risk of a Type 1 error (i.e. rejecting the null hypothesis when it is true) has very important consequences. For example, you might use the 0.01 level when assessing a new filter material for reducing the airborne concentration of hazardous particles. You would need to be reasonably confident that a new material was better than existing ones before recommending it as a replacement.

7.12 (1)   For a population of fossil shells with a mean length of 100 mm and a standard deviation of 10 mm, the finding of a 78 mm shell is unlikely (because it is more than 1.96 standard deviations away from the mean) but not impossible: 5% of individuals in the population would be expected to have shells either ≥ 119.6 mm or ≤ 80.4 mm.

7.12 (2)   The variance calculated from a sample is corrected by dividing by $n - 1$ and not n in an attempt to give a realistic indication of the variance of the population from which it has been taken, because a small sample is unlikely to include sampling units from the most extreme upper and lower tails of the population that will nevertheless make a large contribution to the population variance.

8.10 (1)   These data are suitable for analysis with a paired-sample t test because the two samples are related (the same 10 crystals are in each). The test would be two-tailed because the alternate hypothesis is non-directional (it specifies that opacity may change). The test gives a significant result ($t_9 = 3.161$, $P < 0.05$).

8.10 (2)   The t statistic obtained for this inappropriate independent sample t test is -0.094 and is not significant at the two-tailed 5% level. The lack of significance for this test is because the variation *within* each of the two samples is considerable and has obscured the small but relatively consistent increase in opacity resulting from the treatment. This result emphasizes the importance of choosing an appropriate test for the experimental design.

8.10 (3)   This exercise will initially give a t statistic of zero and a probability of 1.0, meaning that the likelihood of this difference or greater between the sample mean and the expected value is 100%. As the sample mean becomes increasingly different to the expected mean the value of t will increase and the probability of the difference will decrease and eventually be less than 5%.

9.8 (1)   A non-significant result in a statistical test may not necessarily be correct because there is always a risk of *either* Type 1 error *or* Type 2 error. Small sample size will particularly increase the risk of Type 2 error – rejecting the alternate hypothesis when it is correct.

9.8 (2) A significant result and therefore rejection of the null hypothesis following an experiment with only 10% power may still occur, even though the probability of Type 1 error is relatively low.

10.7 (1) (a) The **within group** (error) sum of squares will not be zero, because there is variation within each treatment. (b) The **among group** sum of squares and mean square values will be zero, because the three cell means are the same. (c) A single-factor ANOVA should give $F_{2,9}$ (for treatment) of 0.00. (d) When the data for one treatment group are changed to 21, 22, 23 and 24 the ANOVA should give $F_{2,9}$ (treatment) of 320.0 which is highly significant ($P < 0.001$). (e) The within group (error) mean squares will be the same (1.667 in both cases) because there is still the same amount of variation within each treatment (the variance for the treatment group containing 21, 22, 23 and 24 is the same as the variance within the groups containing 1, 2, 3 and 4).

10.7 (2) (a) Model II – three lakes are selected as random representatives of the total of 21. (b) Model I – the three lakes are specifically being compared. (c) Model I – the six wells are being examined to see whether any give significantly higher yields.

10.7 (3) Disagree. Although the calculations for the ANOVA are the same, a significant Model I ANOVA is usually followed by a posteriori testing to identify which treatments differ significantly from each other. In contrast, a Model II ANOVA is not followed by a posteriori testing because the question being asked is more general and the treatments are randomly chosen as representatives of all possible ones.

10.7 (4) This is true. An F ratio of 0.99 can never be significant because it is slightly less than 1.00 which is the value expected if there is no effect of treatment. For ANOVA, critical values of F are numbers greater than 1.00, with the actual significant value dependent on the number of degrees of freedom.

11.6 (1) (a) Yes, $F_{2,21} = 7.894$, $P < 0.05$. (b) Yes, a posteriori testing is needed. A Tukey test shows that well RVB2 is yielding significantly more oil than the other two, which do not differ significantly from each other.

11.6 (2) An a priori comparison between wells RVB1 and RVB3 using a t test showed no significant difference: $t_{14} = -0.066$, NS. This result is consistent with the Tukey test in 11.6(1).

12.9 (1) (a) For this contrived example where all cell means are the same, Factor A: $F_{2,18} = 0.0$, NS; Factor B: $F_{1,18} = 0.0$, NS; Interaction $F_{2,18} = 0.0$, NS. (b) This is quite difficult and drawing a rough graph showing the cell means for each treatment combination is likely to help. One solution is to increase every value within the three B2 treatments by 10 units, thereby making each cell with B2: 11, 12, 13, 14. This will give Factor A: $F_{2,18} = 0.0$, Factor B: $F_{1,18} = 360.0$, $P < 0.001$, Interaction: $F_{2,18} = 0.0$, NS. (c) Here too a graph of cells mean will help. One solution is to change the data to the following, which, when graphed (with Factor A on the X axis, the value for the means on the Y axis and the two levels of Factor B indicated as separate lines as in Figure 12.1) show why there is no interaction:

Factor A	A1		A2		A3	
Factor B	B1	B2	B1	B2	B1	B2
	1	11	11	21	21	31
	2	12	12	22	22	32
	3	13	13	23	23	33
	4	14	14	24	24	34

12.9 (2) Here you need a significant effect of Factor A and Factor B as well as a significant interaction. One easy solution is to grossly increase the values for one cell only (e.g. by making cell A3/B2 (the one on the far right on the table above) 61, 62, 63 and 64.

13.8 (1) Transformations can reduce heteroscadasticity, make a skewed distribution more symmetrical and reduce the truncation of distributions at the lower and upper ends of fixed ranges such as percentages.

13.8 (2) (a) Yes, the variance is very different among treatments and the ratio of the largest to smallest is 9 : 1 (15.0 : 1.67), which is more than the maximum recommended ratio of 4 : 1. A square root transformation reduces the ratio to 2.7 : 1 (0.286 : 0.106).

14.8 (1) (a) There is a significant effect of distance: $F_{2,8} = 8.267$, $P < 0.05$; and of depth: $F_{4,8} = 3935.1$, $P < 0.001$. (b) When analyzed by single-factor ANOVA, ignoring depth, there is no significant difference among the three cores: $F_{2,12} = 0.016$, NS. The variation among depths within each core has obscured the difference among the three cores, so the researcher would mistakenly conclude there was no significant difference in the concentrations of PAHs and distance from the refinery.

14.8 (2) The glaciologist is using the wrong analysis because the design has locations nested within lakes, so a nested ANOVA is appropriate.

15.8 (1) (a) "....can be predicted from......". (b) "...varies with....."

15.8 (2) (a) The value of r is −0.045, NS. (b) You need to do this by having Y increasing as X increases. (c) You need to do this by having Y decreasing as X increases.

16.13 (1) (a) For this contrived case $r^2 = 0.000$. The slope of the regression is not significant: the ANOVA for the slope gives $F_{1,7} = 0.0$. The intercept is significantly different from zero: $t_7 = 20.49$, $P < 0.001$. (b) The data can be modified to give an intercept of 20 and a zero slope by increasing each value of Y by 10. (c) Data with a significant negative slope need to have the value of Y decreasing as X increases.

16.13 (2) (a) $r^2 = 0.995$. The relationship is significant (ANOVA of slope: $F_{1,7} = 13000.35$, $P < 0.001$) and the regression equation is: weight of gold recovered = −0.002 + 0.024 × volume of gravel processed. The intercept does not differ significantly from zero (and would not be expected to because if no gravel is processed no gold will be recovered).

18.9 (1) This is highly significant ($\chi^2_1 = 106.8$, $P < 0.001$). Because students were assigned to groups at random it seems this high proportion of left-handers occurred by chance, so the significant result appears to be an example of Type 1 error.

18.9 (2) (a) The value of chi-square will be zero. (b) The value of chi-square will increase, and the probability will decrease.

18.9 (3) This is not appropriate because the numbers of trilobites in the two outcrops are independent of each other. The numbers are not mutually exclusive or contingent between outcrops.

18.9 (4) No. The experiment of adding jetties lacked a control for time: the accretion pattern may have changed from the first to the second year simply by chance or as a result of some other factor. It would be helpful to have a control for time where a similar but unmodified coastline was monitored. Often this is not possible, so another approach is to analyze data for the sites over several years prior to the change (i.e. jetty construction) and see if the year(s) after differ significantly from this longer-term data set.

19.10 (1) (a) The relative frequencies are 0.15, 0.30, 0.14, 0.11, 0.03, 0.11, 0.14 and 0.020 respectively. The cumulative frequencies are: 0.15, 0.45, 0.59, 0.7, 0.73, 0.84, 0.98 and 1.0 respectively. (b) For a sample of 100 wells, one distribution of water table depths that would (definitely) not be significantly different to the population would be the numbers in each size frequency division of the population of 1000 wells divided by 10. (c) For a sample of 100 that you would expect to be significantly deeper than the population, you would need to have a much greater proportion in the lower depths. (d) You could use a Kolmogorov–Smirnov test to compare the distributions of the two samples to the population.

19.10 (2) The rank sums are; Group 1: 85, Group 2: 86. (b) There is no significant difference between the two samples: Mann–Whitney $U = 40.0$, NS. (c) One possible change to the data that gives a significant result is to increase the value of every datum in Group 2 by 20.

19.10 (3) One sample appears to be bimodal and there is a gross difference in variance between the two samples. One solution is to transform the data to a nominal scale by expressing them as the number of observations within the two mutually exclusive categories of $\leq 2\,mm$ and $> 2\,mm$. This will give a 2×2 table (Sample 1: 20 individuals are $\leq 2\,mm$, and 2 are $> 2\,mm$; Sample 2: 6 are $\leq 2\,mm$ and 17 are $> 2\,mm$) that can be analyzed using chi-square ($\chi^2_1 = 19.37$, $P < 0.001$; Yates' corrected $\chi^2_1 = 16.80$, $P < 0.001$).

20.19 (1) If there are no correlations within a multivariate data set then principal components analysis will show that for the variables measured there appears to be little separation among objects. This finding can be useful in the same way that a "negative" result of hypothesis testing still improves our understanding of the natural world.

20.19 (2) Eigenvalues that explain more than 10% of variation are usually used in a graphical display, so components 1–3 would be used.

20.19 (3) "Stress" in the context of a two-dimensional summary of the results from a multidimensional scaling analysis indicates how objects from a multidimensional space equal to the number of variables will actually fit into a two-dimensional plane and still be appropriate distances apart. As stress increases it means the two-dimensional plane has to be distorted more and more to accommodate the objects at their "true" distances from each other.

20.19 (4) The "groups" produced by cluster analysis are artificial divisions of continuous data into categories based upon percentage similarity and therefore may not correspond to true nominal categories or states.

21.16 (1) There is no significant long-term linear trend (regression analysis: $F_{1,46} = 0.89$, NS) and a graph shows no obvious repetition. The only significant autocorrelation is an isolated case at lag 10, which suggests little within-sequence repetition or similarity.

21.16 (2) For a relatively short sequence autocorrelations are unreliable because sample size is small. As lag increases the effective sample size is reduced (as less and less of the sequence overlaps with itself), so significant autocorrelations at high lags may be artifacts of having a small number of values in the overlapping section.

21.16 (3) (a) The data could be summarized as the number of heatwaves every 10 years (e.g. 1850–9 etc.) giving 7, 2, 3, 4, 5, 4, 4, 4, 3, 5, 4, 5, 4, 6, 3 and 2 heatwaves for each 10-year interval. The slope of the regression line of the number of summers with heatwaves versus time shows no significant temporal change in the frequency of heatwaves ($F_{1,14} = 0.28$, NS). (c) The frequency distribution of the number of years elapsing between successive heatwaves is: 1yr (23), 2yr (19), 3yr (8), 4yr (5), 5yr (6), 6yr (0), 7yr (1), 8yr (1), 9yr (1). To assess whether years with heatwaves occur at random you need to graph the logarithm of the percentage of "survivors", versus intervals between heatwaves in years (see Table 21.8 for an example). The graph is almost a straight line, so the occurrence of heatwave years appears to be random.

22.6 (1) (a) The expected numbers (in brackets) of quadrats with silver mines is: zero mines (50), 1 mine (69), 2 (48), 3 (22), 4 (8), 5 and more mines (2). $\chi^2_4 = 12.58$ which is just significant at $P < 0.05$. (b) The variance of the number of outcrops per quadrat is 1.34232, so the variance to mean ratio is 0.9726895. (c) This ratio is not significantly different to 1.0: $t_{199} = (0.9726895 - 1)/ 0.08528 = 0.32$, NS. (d) In summary, the result of the chi-square test is consistent with the distribution of silver mines being non-random, but this is only just significant at $P < 0.05$. The t test for uniformity or clustering is not significant although the variance to mean ratio suggests there is a tendency towards uniformity. This example was deliberately chosen to illustrate that these two tests will often give different results when the departure from randomness is slight.

22.6 (2)   (a) For a rose diagram divided into six segments of 60° (1–60° etc.) the numbers per segment are zero, 6, 15, 8, 3 and zero. (b) By inspection the directions of the footprints do not appear to occur equally often. (c) This was confirmed by a chi-square test (expected numbers per segment were 5.33), $\chi_5^2 = 30.64$, $P < 0.01$.

22.6 (3)   (a) For a rose diagram divided into six segments of 60° (1–60° etc.) the numbers of weekly wind direction averages per segment are: 11, 6, 14, 7, 8, and 6. (b) By inspection of these data there is not a great deal of difference in the number of weeks within each segment. (c) A chi-square test (with expected numbers per segment of 8.66) gives $\chi_5^2 = 5.93$, NS. The distribution does not differ significantly among segments, so no advice can be given about the best location of a wind break.

22.6 (4)   The theoretical semivariogram shows the distribution of sampling points and the empirical semivariogram shows a smoothed curve fitted through these.

22.6 (5)   If there is no regional dependence the limits of the semivariance will be the same at any distance from each known point. Therefore, Kriging will only indicate that the value for any predicted point will lie within the expected range of the population. In contrast, if there is regional dependence the relatively narrow limits of the semivariance close to known points will constrain the predicted value.

References

Borradaile, G. J. (2003) *Statistics of Earth Science Data*. New York: Springer.

Chalmers, A. F. (1999) *What Is This Thing Called Science?* 3rd edition. Indianapolis: Hackett Publishing Co.

Davis, J. C. (2002) *Statistics and Data Analysis in Geology*. New York: Wiley.

Fisher, R. A. (1954) *Statistical Methods for Research Workers*. Edinburgh: Oliver and Boyd.

Gamst, G., Meyers, L. S. D. and Guarino, Q. J. (2008) *Analysis of Variance Designs: A Conceptual and Computational Approach with SPSS and SAS*. Cambridge: Cambridge University Press.

Hurlbert, S. J. (1984) Pseudoreplication and the design of ecological field experiments. *Ecological Monographs* **54**: 187–211.

Koch, G. S. and Link, R. F. (2002) *Statistical Analysis of Geological Data*. New York: Dover Publications.

Krumbein, W. C. (1939) Preferred orientation of pebbles in sedimentary deposits. *Journal of Geology* **47**: 673–706.

Kuhn, T. S. (1970) *The Structure of Scientific Revolutions*. 2nd edition. Chicago: University of Chicago Press.

LaFollette, M. C. (1992) *Stealing into Print. Fraud, Plagiarism and Misconduct in Scientific Publishing*. Berkeley, CA: University of California Press.

Lakatos, I. (1978) *The Methodology of Scientific Research Programmes*. New York: Cambridge University Press.

Murray, R. C. and Tedrow, J. C. F. (1992) *Forensic Geology*. Englewood Cliffs, NJ: Prentice Hall.

Popper, K. R. (1968) *The Logic of Scientific Discovery*. London: Hutchinson.

Siegel, S. and Castallan, J. J. (1988) *Statistics for the Behavioral Sciences*. 2nd edition. New York: McGraw-Hill.

Singer, P. (1992) *Practical Ethics*. Cambridge: Cambridge University Press.

Sokal, R. R. and Rohlf, F. J. (1995) *Biometry*. 3rd edition. New York: W. H. Freeman.

Sprent, P. (1993) *Applied Nonparametric Statistical Methods*. 2nd edition. London: Chapman & Hall.

Stanley, C. R. (2006) Numerical transformation of geochemical data: 1. Maximizing geochemical contrast to facilitate information extraction and improve data presentation. *Geochemistry: Exploration, Environment, Analysis* **6**: 69–78.

Strom, R. G., Schaber, G. G. and Dawson, D. D. (1994) The global resurfacing of Venus. *Journal of Geophysical Research* **99**: 10899–10926.

Student (1908) The probable error of a mean. *Biometrica* **6**: 1–25.

Tukey, J. W. (1977) *Exploratory Data Analysis*. Reading: Addison-Wesley.

Ufnar, D. F., Groecke, D. R. and Beddows, P. A. (2008) Assessing pedogenic calcite stable isotope values; can positive linear covariant trends be used to quantify palaeo-evaporation rates? *Chemical Geology* **256**: 46–51.

Vermeij, G. J. (1978) *Biogeography and Adaptation: Patterns of Marine Life*. Cambridge, MA: Harvard University Press.

Woods, S. C., Mackwell, S. and Dyar, D. (2000) Hydrogen in diopside: Diffusion profiles. *American Mineralogist* **85**: 480–487.

Zar, J. H. (1996) *Biostatistical Analysis*. 3rd edition. Upper Saddle River, NJ: Prentice Hall.

Zelen, M. and Severo, N. C. (1964) Probability functions. In *Handbook of Mathematical Functions*, Abramowitz, M. and Stegun, I. (eds.). Washington, DC: National Bureau of Standards, pp. 925–995

Index